Production Technology on Bio-Organic Farm Inputs

A.K. Singh, Ph.D.
Associate Professor (Agriculture)
North Eastern Regional Institute of Water and Land Management
Tezpur, Assam (India)
Member, International Competence Centre for Organic Agriculture
Associate Trustee, Indian Organic Certification Agency

International Book Distributing Co.
(Publishing Division)

Published by

INTERNATIONAL BOOK DISTRIBUTING CO.
(Publishing Division)
Khushnuma Complex Basement
7, Meerabai Marg (Behind Jawahar Bhawan),
Lucknow 226 001 U.P. (INDIA)
Tel. : 91-522-2209542, 2209543, 2209544, 2209545
Fax : 0522-4045308
E-Mail : ibdco@airtelbroadband.in

First Edition 2008

Price:

ISBN 81-8189-216-X

Composed & Designed at :

Panacea Computers
3rd Floor, Agarwal Sabha Bhawan, Subhash Mohal
Sadar Cantt., Lucknow-226 002
Phone : 0522-2483312, 9335927082
E-mail : prasgupt@rediffmail.com

Printed at:

Salasar Imaging Systems
C-7/5, Lawrence Road Industrial Area
Delhi - 110 035
Tel. : 011-27185653, 9810064311

Dr. R.B. Srivastava
Director

Defence Research Laboratory,
Tezpur, Assam
DRDO, Ministry of Defence,
Govt. of India

Foreword

Organic agriculture has a greater sequestration potential as it follows the key principle of tight nutrient and energy cycle through organic matter management in soils. This is achieved through improved practices in cropland management using organic seeds/plants and bio-inputs for nutrient, pest and diseases management.

In view of the immense potential of organic farming in the country, this book is compiled to provide a convenient and concise source of information on production techniques of Compost, Vermicompost, Biofertilizers, BGA and Mycorrhizae, Biopesticides and Organic Nursery Development using tissue culture techniques. Information on Soil Management and Nutrition, Organic Pest and Disease Management and Organic Input Evaluation is also incorporated.

I hope that this technical book will enhance the knowledge on production of organic inputs in the area delineated for "Organic Farming". It may also contribute to generate new and sustainable incomes for many small farmers and also a good business for all the links in the supply chain towards the emerging organic markets.

11th October, 2007

(Dr. R.B. Srivastava)

Preface

Organic farming is an environmental friendly ecosystem management in which, use of all kinds of synthetic inputs are eliminated. The area delineated for "organic farming" forbids use of synthetic fertilizers, pesticides, genetically modified seeds (GMO) and breeds, etc. These are replaced with site-specific management system that maintain and increase long-term soil fertility and prevent pest and diseases.

Organic farming uses environmental friendly inputs like compost, vermicompost, biofertilizers, biopesticides, organically grown seeds and seedlings and therefore generates environmental friendly foods and services. Such farming, therefore, positively contributes to marked reduction in air, soil and groundwater pollution. Moreover, it is a solution to the problem created by nitrate pollution and pesticide residues, improves soil structure, fertility and soil fauna.

This technical guidebook presents production techniques of compost, vermicompost, biofertilizers, biopesticides and organic nursery. The book is useful for organic operators, students, NGOs planners, trainers, agriculture extension officers, progressive farmers, entrepreneurs, consultants, libraries and others actively involved in development of organic farming.

Dated: 10th October, 2007

A. K. Singh
Associate Professor (Agri),
NERIWALM, Tezpur, India

Contents

Foreword
Preface

I	**SOIL MANAGEMENT AND NUTRITION**	**1-22**
	1. Rational Use of Agro-ecosystem	3
	1.1 The organic matter	3
	1.2 The soil-microcosm	3
	1.3 Nitrogen fixing plants	4
	1.4 Green manures and cover crops	5
	1.5 Associating crops and crop rotation	9
	1.6 Biomass production	11
	1.7 Utilization of on-farm resources	11
	2. Suitable Soil Management Practices	14
	2.1 Mulching	13
	2.2 Soil cultivation and tillage	15
	2.3 Soil compaction	17
	2.4 Soil erosion	17
	3. Plant Nutrition and Plant Health	18
	3.1 Nutrient supply	19
	3.2 Nutrient recycling	19
	3.3 Fertility improvement	22
II	**COMPOSTING**	**23-36**
	1. Process of Composting	23
	1.1 Selecting the primary materials	23
	1.2 Setting up a compost heap	25
	2. Systems of Composting	26
	2.1 Indore method	26
	2.2 Banglore method	28
	2.3 VAT Method	29
	2.4 NADEP method	31
	2.5 Water hyacinth compost	34
	2.6 Compost Tea	35

III VERMICOMPOSTING 37-50
1. About the Worms 37
2. Requirements for Vermicomposting 38
3. Production of Vermicompost 41
3.1 Method of vermicomposting 41
3.2 Vermicast collection 43
3.3 Vermiculture 44
3.4 Vermiwash 45
4. Costs of Production 46
5. Advantage of Vermicompost on Organic Farm 49

IV BIOFERTILIZER PRODUCTION 51-72
1. Biofertilizer and their Beneficiary Crops 51
2. Biofertilzer Production Technology 52
2.1 Quality standards of biofertilizers 54
2.2 Requirements of biofertilzer production 54
3. Commercial Production of Biofertilizers 56
3.1 Production techniques for Rhizobium 60
3.2 Production techniques for Azotobacter 63
3.3 Production techniques for Azospirillum 64
3.4 Production techniques for Phosphobacterin 66
4. Economics of the Biofertilizer Production 67
5. Technical Aspects of Biofertilizer 69
5.1 Methods of placement of biofertilizers 69
5.2 Mode of action 70
5.3 Critical factors for effectiveness of iofertilizers 70
5.4 Level of benefits 71

V AZOLLA, BGA & MYCORRHIZA (VAM) PRODUCTION 73-81
1. Production Technique of Azolla 74
1.1 Ideal conditions for growth and multiplication of Azolla 74
1.2 Method of growing Azolla in nursery 74

2. Production Techniques of Blue Green Algae (BGA) 75
2.1 Mass Production technique 77
3. Field Inoculation and Multiplication of VA Mycorrhizae 78
3.1 Type of Mycorrhizae 78
3.2 Inoculation methods of Mycorrhizae production 79

VI PLANT DISEASE AND PEST MANAGEMENT 83-91
1. Preventive Measures for Disease and Pest Management 84
2. Curative Crop Protection Methods 86
3. Use of Bio-Fungicides and Bio-Pesticides 88
4. Other Methods of Pest and Disease Control 90

VII BIO-PESTICIDE PRODUCTION 93-127
1. Commercial Production of Bio-pesticides 94
1.1 Basic requirements 94
1.2 Scale of production 95
1.3 Bio-agents for commercial production 97
1.4 Biopesticide production process 98
2. Bio-control Units for Insect-Pests 99
2.1 Production technology of Trichogramma egg parasite 99
2.2 Production technology of Chrysopid predetors 103
2.3 Production technology of Australian ladybird beetle 105
2.4 Production technology of NPV 106
2.5 Sex pheromone traps 116
3. Bio-control Units for Plant Disease 117
3.1 Production technology of Trichoderma fungi 117

4. Pesticides Production Unit 119
4.1 Unit Size 119
4.2 Techno-economic parameters 119
5. Bordeaux Mixture 125
6. Non-Pesticidal Preparation 126
6.1 Neem seed kernel suspension 126
6.2 Chilli-garlic extract 126
6.3 Cattle-dung and urine extract 127
7. Technical Consultancy 127

VIII NURSERY DEVELOPMENT 129-147
1. Nursery Technique 130
1.1 Cost of cultivation 131
2. Plant Propagation by Tissue Culture 132
2.1 Requirements for tissue culture plant propagation 134
2.2 Unit size 140
2.3 Estimated production, cost and income 140
2.4 Tissue culture technology 141

IX INPUT EVALUATION 149-154
1. Evaluation of Organic Inputs 149
2. How to apply for input approval 152
3. Information required for Input Approval 153

Frequently Asked Questions 155-169

References 171

Appendices 173-182

Index 183-184

Chapter 1

Soil Management and Nutrition

Soil is the most important production factor for crops and at the same time is also the most influenced by the farmer. Soils are very diverse and complex systems with full of life. The soil itself can be viewed as a living organism, because it is a habitat for plants, animals and microorganisms that are all interlinked.

Soil consists in mineral particles, organic matter and pores. Mineral particles originate from subsoil and rock, which gets crushed to smaller and smaller pieces (sand, silt and clay) through physical and chemical weathering processes. Mineral particles contain nutrients that are slowly released in the process of weathering. Plant roots and some microorganisms can actively dissolve nutrients from mineral salts and use them for their growth. In addition to mineral salts, soil contains organic matter, resulting from the decomposition of biomass. Organic matter is mainly present in the top layer of the soil, which is subject to continuous transformation processes. Soil organic matter can be further decomposed by soil organisms. The resulting structures can recombine to form very stable humus structures, which can remain in the soil for many years and contribute significantly to the improvement of the soil structure.

What do organic standards say on plant nutrition?

The approach to plant nutrition in organic agriculture is fundamentally different from the practices of conventional agriculture. While conventional agriculture aims at providing direct nutrition to the plants by using mostly easily soluble chemical fertilizers, organic farming feeds the plants indirectly by feeding the soil organisms with organic matter. Organic soil fertility management is based on rational use of native resources achieved through crop rotation, cultivation of legumes, green manures or deep rooting plants and reutilization of organic farm by-products. Use of auxiliary resources in soil fertility management, i.e. mineral fertilizers and soil improvers that are not obtained directly from the agro-ecologic system involved and are acquired on the market, should only be employed as a second choice.

IFOAM Basic Standards as well as national regulations (NSOP) define how plant nutrition should be approached in organic agriculture and which materials are allowed, with restrictions and which are prohibited. Following nutrition management practices are recommended:

- Biodegradable material builds the basis of the fertilization program.
- The total amount of biodegradable material brought onto the farm unit is limited.
- Animal runs should be prevented from becoming over-manured where there is a risk of pollution to rivers or groundwater.
- Brought-in material shall be in accordance with a positive list (list of allowed fertilizers-Appendix-1).
- No manures containing human excrements can be used as fertilizer on vegetation for human consumption if not first sanitized.
- No chemical fertilizers containing nitrogen can be used; Chilean nitrate and all synthetic nitrogenous fertilizers, including urea, are prohibited.
- Restricted use of chemical magnesium and trace elements and/or fertilizers with unwanted substances, e.g. basic slag, rock phosphate and sewage sludge. Chemical magnesium and trace elements shall be used only after soil analysis, with prior permission of the certifier and as a supplement to organic sources.

1. Rational Use of Agro-ecosystem

1.1 The organic matter

Organic matter has a great capacity to retain nutrients and release them continuously (nutrient exchange capacity). It thereby increases the capacity of the soil to supply the plants with nutrients and reduces nutrient losses by leaching. This is especially important in ferralitic and sandy soils as they naturally retain very few nutrients. Organic matter also prevents soils from becoming too acidic.

Soil organic matter helps to build up a loose and soft soil structure with a lot of pores. This leads to better aeration, better infiltration of water and an easier penetration of roots.

The visible parts of organic matter act like tiny sponges which can hold water up to five times their own weight. Therefore, in dry periods more water is available for the plants for a longer time. This is especially importantin sandy soils. The non-visible parts of organic matter act like a glue, sticking soil particles together, thus forming stable crumbs. Such aggregates improve the soil structure, especially in clay and sandy soils.

Beneficial microorganisms and other soil organisms such as earthworms also feed on organic material, thus decomposing it. As these organisms require sufficient humidity and aeration, soil organic matter provides a suitable environment for them.

1.2 The soil-microcosm

A teaspoon of active soil is the habitat of millions of soil organisms. Some are of animal origin and some are of plant origin. The organisms vary greatly in size. Some are visible to the naked eye, such as earthworms, mites, termites, etc. Most of them, however, are so small that they can only be seen with a microscope, thus they are called microorganisms. The most important microorganisms are bacteria, fungi and protozoa. Microorganisms are the key elements to the quality and fertility of soils, they do their work invisibly. The greater the variety of species and the higher their number, the greater the natural fertility of the soil.

Soil organisms are important because they:

- help to decompose organic material and build up humus
- mingle organic matter with soil particles and thus help to build stable crumbs
- dig tunnels, which encourages deep rooting of plants and good aeration of the soil
- help to release nutrients from mineral particles
- control pest and disease organisms affecting the roots of crops.

As the plant roots and the soil organisms consume air, good air circulation within the soil is crucial for their development. Soil organism activity is generally low when soils are dry, very wet or too hot. Activity is highest in warm, moist soils when food (i.e. biomass) is available. Earthworms accelerate the decomposition of biomass by removing dead plant material from the soil surface.

1.3 Nitrogen fixing plants

Air offers potentially endless amounts of nitrogen. Plants of the legume and mimosa family are capable of fixing nitrogen from the air with their roots to use as a nutrient. Legumes do this by living in association (symbiosis) with bacteria called rhizobium that are hosted in nodules growing on the roots. These bacteria take nitrogen from the air, transform it and make it available for the host plant. Bacteria take the necessary energy from the plant roots (sugars, the products of photosynthesis). The blue-green algae, e.g. "azolla" growing in rice fields, produce the energy through their own photosynthesis. The partnership between plant and rhizobia is usually very specific. For this reason, it may be necessary to introduce (inoculate) the bacteria legume plants which are grown in a field. The better the nutrient and water supply, soil qualities including soil acidity, temperature and light for the plant, the better the legume can supply the bacteria with energy and satisfy its own nitrogen needs. Among nitrogen fixing plants the annual and the perennial species can be distinguished. In alley cropping, perennial shrubs are grown in rows between the main crop.

The benefits of nitrogen fixing trees and shrubs are:

- The leaves and twigs of nitrogen fixing trees are rich in nitrogen and other plant nutrients and are a valuable free source of fertilizer. With their roots, they directly increase the nitrogen content of the soil and build up soil organic matter.
- *Wood and timber*: Some luxury timbers are provided by nitrogen fixing trees. Fast-growing nitrogen fixing trees also produce excellent fuel wood and charcoal.
- *Fodder and food*: The highly nutritious and digestible leaves of some nitrogen fixing trees make them excellent feed for animals. Several species of nitrogen fixing trees produce food for humans (e.g. drumstick, tamarind).
- *Protection and support:* Nitrogen fixing trees can be grown as living fences and hedges to protect crops from wildlife, domestic animals, and people. Trees with dense canopies can be grown as a windbreak or to protect organic farms from conventional neighbours. Nitrogen fixing trees may be grown to provide shade for tea, cacao or coffee or to provide support for climbing crops such as yams, vanilla and black pepper.

1.4 Green manures and cover crops

The main aim of green manures is to provide nutrients to subsequent crops and to increase soil fertility through addition of organic matter. Cover crops have similar benefits as green manure, and in many cases the same crops and management methods are used.

There is a way to distinguish cover crops from green manure:

- Cover crops are in most cases perennial and not incorporated into the soil. After cutting, plant material is left on the soil surface or harvested as animal fodder or compost material.

- Green manures are mostly temporary. They are worked into the soil, where the fresh plant material releases nutrients quickly and will be fully decomposed within a short period of time.

Cover crops and green manures have a number of benefits:

- They penetrate the soil with their roots, make it more friable and to bind nutrients that would otherwise be washed away.
- They suppress weeds and protect the soil from erosion and direct sunlight.
- If leguminous plants are used, nitrogen is fixed from the air into the soil.
- Some green manures and cover crops can be used as fodder plants or even to provide food for human consumption (e.g. beans and peas).
- By decomposing, green manures and cover crops release all kinds of nutrients for the main crops to utilize, thus improving their yield.
- The incorporated plant material builds up organic matter in the soil and activates soil organisms. This improves soil structure and water holding capacity.

The following aspects must be considered before growing green manures and cover crops:

- Labour is required for tillage, sowing, cutting and incorporation of plants into the soil, and is most intensive where the amount of helpful equipment available is limited.
- If green manures and cover crops are intercropped with the main crops, they might compete for nutrients, water and light.
- When old or coarse plant material is incorporated into the soil, nitrogen may be temporarily immobilized and therefore unavailable for plant growth (nitrogen immobilization).

- If food and space are in short supply, it may be more appropriate to grow a food crop rather than a green manure and recycle the crop residues, or to intercrop a green manure crop with the main crop.
- The benefits of green manures and cover crops occur over the long term and are not always immediately visible.

1.4.1 Sowing the green manure and cover crops

- If grown within a crop rotation, the sowing time must be chosen to enable the green manure to be cut down and worked into the soil before the next crop is sown.
- Green manures and cover crops need water for germination and growth.
- The ideal seed density must be tested for each individual situation.
- In general, no additional fertilization is necessary. If legumes are grown in a field for the first time, inoculation of the seeds with the specific rhizobia may be necessary to profit from nitrogen fixation of the legume.
- If under-sown, the green manure is sown at the same time as the main crop. If it grows faster than the main crop and competition is too high, it can also be sown later when the crop has been established. Later sowing may be combined with a weeding passage.

1.4.2 Green manure into the soil

Timing: The time gap between digging in the green manure and planting the next crop should not be longer than 2 to 3 weeks so as to prevent nutrient losses from the decomposing green manure.

Crushing: Green manures are worked easily when the plants are still young and fresh. If the green manure plants are tall or contain bulky and hard plant parts, it is preferable to chop the plants into pieces to allow for easier decomposition. The older the plants, the longer decomposition will take. The best time to dig in green manure plants is just before flowering.

Depth of incorporation: Green manures should not be ploughed deeply into the soil. Instead, they should only be worked in to the surface soil (in heavy soils only 5 to 15 cm deep, in light soils 10 to maximum 20 cm deep). In warm and humid climates the material can also be left on the soil surface as a mulch layer.

1.4.3 Selection of right species

There are a large variety of plants, especially legumes that can be used as green manure crops. The following characteristics make an ideal green manure or cover crop:

- The seeds are cheap, easy to get, to harvest, to store and to propagate.
- Adapted to the local growing conditions.
- Fits into the crop rotation or fits with the main crop (e.g. fruit trees, coffee, cocoa, etc.).
- Possesses a rapid growth rate and be able to cover the soil in short time.
- Resistant to pests and diseases.
- Competitive with undesired spontaneous vegetation (e.g. aggressive grasses).
- Does not pose a risk of transmitting diseases and pests to other crops.
- Produces large amounts of organic matter and dry material.
- Fixes nitrogen from the air and provide it to the soil.
- De-compacting root system and regenerate degraded soils.
- Easy to sow and to manage a single crop or associated with other crops.
- Can be used as fodder, and food grains.

An alternative to sowing a green manure or cover crop in the field is to collect fresh plant material from elsewhere and work it into the soil as mulch.

1.5 Associating crops and crop rotation

In many traditional agricultural systems, a diversity of crops in time or space can be found. There are different reasons why farmers do rotate or associate crops. Different plant species respond to the characteristics of the soil, have different root systems and have different needs for nutrients, water, light, temperature and air.

1.5.1 Associating crops

Associating crops is defined as the growing of two or more crops in the same field at the same time. If suitable crops are combined, mixed cultivation can lead to a higher total yield per area. This is basically due to the more efficient use of space (over and under ground) and because of beneficial interactions between the mixed crops. A greater diversity of crops can be grown in the fields. This helps the farmer to avoid dependence on only one crop, ideally achieving a continuous supply of products from the field. Associating crops have agro-ecological benefits like:

- The diversity makes it more difficult for pests and germs to attack a certain species.
- Mixed cropping with legumes improves nitrogen supply of non-legumes.
- Associated crops cover the soil faster and grow more densely, thus suppressing weeds more efficiently.

There are different possibilities to associate crops:

Mixed cropping: Two or more crops are sown at the same time-sharing the same space, or they are sown at the same time in neighbouring rows. One crop may also be sown as a border crop.

Cropping in lines: Two or more crops are sown at the same time in neighbouring lines with wide spacing.

Relay cropping: A second crop is being sown before the harvest of the first one.

Combined cultivation: Cultivation of trees and annual crops together.

1.5.2 Selection of crops under mixed cropping

Crops and species grown in association should have different growth habits and different needs for light. The selection of crops should be done in light of following principles:

- Crops with strong rooting should be alternated with crops with a weak root growth. Crops with deep rooting are best grown together with species with shallow root growth. The periods of most active nutrient uptake should not coincide.
- Plant distances should be such that nutrient competition between plants can be minimized.
- Perennial plants can be well associated with seasonal plants.
- Leguminous crops may be grownin association with crops or before crops that have a high demand for nitrogen.

1.5.3 Crop rotation

If the same crop is grown for several consecutive years on the same land yields will normally decline (or more fertilizer will be needed to reach the same yield) and health problems will arise in the crop or field. Weeds that are well adapted to the conditions offered by the crop (e.g. good light conditions, typical soil cultivation) may spread and require increased efforts to be controlled.

Benefits of crop rotation:

- When different crops are grown in sequence in the same field, each crop uses the soil in its own particular way and thus reduces the risk of nutrient depletion. A well-balanced alternation of crop species also prevents the development of soil-borne diseases. Therefore, cultivation pauses must be respected for the same crop and among crops of the same plant family.
- To avoid the development of persistent weeds, plants with a slow growth should be grown after crops possessing

good weed suppression. A change between deep and flat rooting crops and between crops building high stalks and species producing a great leaf mass that covers the soil quickly also helps to suppress the weeds.

- Crop rotation is also an important instrument to maintain soil organic matter. Ideally, crop rotation should maintain, or even raise, the content of soil organic matter.

1.6 Biomass production

- Integrate green fallow periods with green manures in the crop rotation.
- Applying compost and animal manures.
- Aim at having the soil covered with plants the whole year round.
- Integrate fodder cultivation in the farm (grass, fodder, hedges).
- Use unproductive space (e.g. along paths, field borders, steep slopes) for planting trees or hedgerows.
- Use unproductive fields and unproductive time in a rotation (in between two crops) for planting vigorous nitrogen fixing crops (such as *Canavalia* spp. and *Cayanus cayan*).
- Establish agroforestry systems where appropriate.
- Leave single trees standing in the field (e.g. nitrogen fixing trees), manage them with intense pruning.

1.7 Utilization of on-farm resources

1.7.1 Compost

Compost is the most important fertilizer in organic agriculture. Organic growers are therefore very much concerned with producing good compost. Composting is the process of transforming organic material of plant or animal origin into humus substances that are relatively resistant to microbial decomposition. Compost thus helps to maintain or increase soil organic matter content. The other components of compost provide nutrients and micronutrients in the right proportion (as compost is built from plant materials) for plants to utilize. Compost has both a long and short term effect on plant nutrition as nutrients are permanently released. Due to its neutral pH, compost improves the availability of nutrients in acid soils. When mixed with soil, compost can suppress soil borne disease pathogens. Mature compost is good for plants and does not impede plant roots and microorganisms in the soil as do substances released during a rotting process.

Compost can be used as soon as the original composting material is not recognizable anymore. The compost has turned a dark brown or blackish colour and has a pleasant smell.

Usually it is not possible to produce sufficient amounts for fertilizing all fields. Therefore, farmers should think carefully about where compost application would be most beneficial. High efficiency is achieved in nurseries and when planting seedlings or saplings. Quantities and timing of application depend on the compost quality and differ from crop to crop. The farmers should take into consideration that during the decomposition process some organic matter and nutrients would be lost. Also, compost production is labour intensive and demands regular attention.

1.7.2 Vermicompost

Vermicompost is mainly based on the activity of worms and does not go through a heating phase at all. Earthworms are highly efficient in transforming biomass such as leaves into excellent humus within a short period of time. During the digestion of organic material, they mix organic and mineral soil particles and build stable crumbs, which help improve the soil structure. They have high nutrient levels and good water retention. The excrements contain 5x more nitrogen, 7x more phosphate, 11x more potash and 2x more magnesia and calcium than normal earth. Their tunnels promote infiltration and drainage of rainwater and thus prevent soil erosion and water logging. In addition, the excrement has a growth promoting effect on plants.

Some experienced farmers use 'vermi-wash', the liquid collected from the compost heap after sprinkling, as a leaf fertilizer and plant tonic. This can even help plants to get rid of pests (e.g. aphids) and diseases. Worms are very sensitive to fluctuations in moisture and temperature.

Earthworms need sufficient supply of biomass i.e. compost material, moderate temperature and sufficient humidity. Frequent tillage decreases the number of earthworms in the soil, as does the use of pesticides. They are also attacked by ants and termites. Therefore, a solid base is needed to protect the worms from predators. Though vermicompost is definitely a very good manure, it requires more investment (tank and worms), labour and permanent care as compared to ordinary composting methods.

2. Suitable Soil Management Practices

2.1 Mulching

Mulching is the process of covering the topsoil with plant material such as leaves, grass and crop residues. A mulch cover enhances the activity of soil organisms such as earthworms. They help to create a soil structure with plenty of smaller and larger pores through which rainwater can infiltrate easily into the soil, thus reducing surface runoff. As the mulch material decomposes, it increases the content of organic matter in the soil. Soil organic matter helps to create good soil with a stable crumb structure. Thus, the soil particles will not be easily carried away by water. Therefore, mulching plays a crucial role in preventing soil erosion.

2.1.1 Selection of mulch materials

The kind of material used for mulching will greatly influence its effect. Material that easily decomposes will protect the soil only for short time, but will provide nutrients to the crops while decomposing. Hardy materials will decompose more slowly and therefore cover the soil for a longer time and protect against erosion. If the decomposition of the mulch material should be accelerated, animal manures may be spread on top of the mulch, thus increasing the nitrogen content.

Sources of mulching material can be the following:

- Cover crops, grasses and weeds
- Crop residues (straw etc.)
- Pruning materials from trees and hedges
- Wastes from agricultural processing or from forestry.

2.1.2 Constraints of mulching

- Slugs, snails, ants or termites may find ideal conditions for living under a mulch layer and can multiply quickly. They may cause damage to the crops.
- When crop residues are used for mulching, there is an increased risk of sustaining pests and diseases.
- Damaging organisms such as stem borers may survive in the stalks of crops like cotton, corn or sugarcane.
- Plant material infected with viral or fungal diseases should not be used if there is a risk that the disease might spread to the next crop.
- When carbon rich materials such as straw or stalks are used for mulching, nitrogen from the soil may be used by microorganisms to decompose the material. Thus, nitrogen may be temporarily unavailable for plant growth if the applied plant material does not contain sufficient nitrogen (risk of N-immobilization).
- The major constraint for mulching is usually the availability of organic material. Its production or collection normally involves labour and may compete with the production of crops.

2.1.3 Application of mulch

- If possible, the mulch should be applied before or at the onset of the rainy season, when the soil is most vulnerable.

- If the layer of mulch is not too thick, seeds or seedlings can be directly sown or planted in between the mulching material. On vegetable plots, it is best to apply mulch only after the young plants have become somewhat hardier, as they may be harmed by the decomposition products from fresh mulch material.

If mulch is applied prior to sowing or planting, the mulch layer should not be too thick in order to allow seedlings to penetrate it. Mulch can also be applied in established crops, and is best down directly after digging the soil. It can be applied between the rows, directly around single plants (especially for tree crops) or evenly spread on the field.

2.2 Soil cultivation and tillage

Careful soil cultivation can improve the soil's capacity to retain water, its aeration, capacity of infiltration, warming up, evaporation etc. But soil cultivation can also harm the soil fertility as it accelerates erosion and the decomposition of humus. Depending on the cropping system and the soil type, appropriate soil cultivation patterns must be developed.

Objectives of soil cultivation:

- Loosen the soil to facilitate the penetration of plant roots.
- Improve the aeration (nitrogen and oxygen from the air).
- Increase infiltration of water.
- Reduce evaporation.
- Encourage the activity of the soil organisms.
- Destroy or control weeds and soil pests.
- Incorporate crop residues and manures into the soil.
- Prepare the site for seeds and seedlings.
- Repair soil compaction caused by previous activities.

2.2.1 Minimum and zero-tillage

Regular tillage accelerates the decomposition of organic matter which can lead to nutrient losses. The mixing of soil layers can severely harm certain soil organisms. Soil after tillage is very prone to soil erosion if left uncovered before the onset of heavy rains. Zero-tillage systems help to build up a natural soil structure with a crumbly top soil rich in organic matter and full of soil organisms. Nutrient losses are reduced to a minimum. Soil erosion won't be a problem as long as there is a permanent plant cover or sufficient input of organic material. Farmers can save a lot of labour. However, zero-tillage is a challenge for organic producers.

Tillage is especially in annual crops, one important tool for weed management and therefore widely practiced in organic agriculture. To minimize the negative impact of soil cultivation while benefiting from its advantages, the organic farmer should aim at reducing the number of interventions to the minimum and choose methods that conserve the natural qualities of the soil.

2.2.2 Types of soil cultivation

Depending on the aim of the soil cultivation, different cultivation practices are implemented during different stages of the cropping cycle:

Post-harvest: In order to accelerate decomposition, the residues of the previous crop are incorporated into the soil (15 to 20 cm) before preparing the seedbed for the next crop.

Primary tillage: In annual crops or new plantations, primary tillage is usually done with a plough or a similar instrument. In principle, soil cultivation should achieve a flat turning of the top soil and a loosening of the medium deep soil.

Seedbed preparation: Before sowing or planting, secondary soil cultivation is done to crush and smoothen the ploughed surface. If weed pressure is high, seedbeds can be prepared early thus allowing weed seeds to germinate before the crop is sown.

In-between the crop: Once the crop is established, shallow soil cultivation is applied to suppress weeds, enhances the aeration of the soil, to reduce the evaporation of soil moisture from the deeper soil layers and to stimulate the decomposition of organic matter thus making nutrients available.

Tools should be chosen considering the soil cultivation purpose, the soil type, the crop and the available power source:

- Primary cultivation: pole plough, mouldboard
- Plough, digging fork, spade
- Secondary cultivation: cultivators, harrows, rakes
- Inter-row cultivation: inter-row cultivators, hoes
- Land forming: ridgers, hoes.

2.3 Soil compaction

If soils are cultivated in wet conditions or burdened with heavy machinery, there is a risk of soil compaction which results in suppressed root growth, reduced aeration and waterlogging. Where soil compaction is a potential problem, farmers should be aware of the following aspects:

- The risk of compaction is highest when the soil structure is disturbed in wet conditions.
- Do not drive vehicles on your land soon after rains.
- Ploughing of wet soils can lead to a smearing of the plough sole.
- Soils rich in sand are less prone to soil compaction than soils rich in clay.
- High content of soil organic matter reduces the risk of soil compaction.
- It is very difficult to restore good soil structure once soil compaction has taken place.
- Deep tillage in dry conditions and the cultivation of deep-rooted plants can help to repair soil compaction.

2.4 Soil erosion

During the dry season, ground vegetation usually becomes scarce and thin, leaving the soil uncovered. As a result, when the rains arrive, large amounts of valuable topsoil can be washed away, leaving the land uneven with gullies and low fertility soil. Not only steep slopes but also plain fields are also prone to soil erosion, and can be severely affected. Besides rain, excessive irrigation can also cause soil erosion.

Strategies for preventing soil erosion should ideally be combined:

1. Reducing the erosive power of the rain drops by keeping the soil covered (with vegetation or mulch). Cropping systems should be designed in such a way that the soil is almost permanently covered with plant canopy.
2. Improving the infiltration of the rainwater into the soil. The way to improve the infiltration is improving the soil structure.
3. Reducing the speed of the water flowing down the slopes with the help of construction such as bunds, stonewalls, living barriers, trenches, terraces, etc.

3. Plant Nutrition and Plant Health

Plant nutrition and plant health are closely linked. Chemical fertilization has the following negative impact on soil and plant health:

- Chemical fertilization reduces the colonization of plant roots with the beneficial root fungus mycorrhiza.
- High nitrogen fertilization stops symbiotic nitrogen fixation by rhizobia.
- Oversupply of nitrogen leads to a softening of the plants' tissues resulting in plants which are more sensitive to diseases and pests.
- The exclusive use of NPK-fertilizers leads to a depletion of micronutrients in the soil, as these are not replaced by such fertilizers. This results in a decline of yields and a reduction in plant and also animal health.
- Decomposition of soil organic matter is enhanced, which leads to a degradation of the soil structure and a higher vulnerability to drought.

3.1 Nutrient supply

Plant nutrition in organic farming focuses on sound management of soil organic matter. The organic farmer uses three approaches to ensure a continuous nutrient supply from soil organic matter:

Varying the input of organic material: The amount and the quality of organic matter influence the content of organic matter in the soil. A regular supply of organic matter provides the best conditions for balanced plant nutrition. Estimates say that in humid tropical climates8.5 tones, in sub-humid climate 4 tones, and in semi-arid 2 tones of biomass is needed per hectare and per year to maintain soil carbon levels of 2, 1 and 0.5 % respectively.

Suitable crop rotation: The crops being grown determine the amount of nutrients the soil needs in order to maintain its fertility. The farmer arranges the rotation in such a way that demand and supply of nutrients (e.g. nitrogen from legumes, nutrients from a green manure crop) fit in the best possible way.

Influencing nutrient mobilization: The farmer can influence the nutrient release from humus by cultivating the soil at the appropriate time, to the appropriate depth, and with the appropriate intensity and frequency. Soil cultivation improves aeration of the soil and enhances the activity of soil microorganisms. If the microorganisms find suitable conditions for their growth, they can be very efficient in dissolving nutrients and making them available to plants. Therefore, in organic agriculture, it is important to encourage plant health by creating biologically active soil.

3.2 Nutrient recycling

Organic growers aim to achieve a more efficient use of farm-own nutrients and to reduce external inputs to a minimum. This idea leads to the concept of closed nutrient cycles. It is clear that the export of nutrients with market goods and losses through leaching and volatilization and erosion cannot be avoided completely. In organic farming, the big question is: 'how to optimize nutrient management on the farm?'

There are three principles of how to optimize nutrient management.

Principle 1: Minimize Losses

- High losses of nutrients result from leaching due to the low exchange capacity of the soil. Leaching can be reduced by raising the content of soil organic matter.
- If dung or compost is kept in waterlogged conditions or is exposed to the sun, high losses of nitrogen may occur. Washout of soluble nutrients from stored dung and compost can be prevented by proper sheltering and storage.
- Dung or compost is often stored in pits where water collects during the rainy season. Nitrogen gets lost through leaching (if the bottom of the pit is permeable) or through volatilization (if the water gets logged in the pit).
- Soil erosion robs the soil of its most fertile part i.e. the top soil, which contains the majority of nutrients and organic material. This can be prevented by maintaining a dense plant cover and with constructions such as terracing.
- Avoid burning biomass.
- To prevent losses of nitrogen fixed by leguminous plants, practice mixed cropping or crop rotation with species of high nitrogen demand.
- Nutrient release from soil organic matter when there are no plants present or able to take it up leads to considerable nutrient losses.
- Nitrogen is easily lost by volatilization. The highest losses occur during the first two hours after manure is applied to the field. Therefore, farmyard manure should be applied in the evening as cool night temperatures and the higher humidity reduce the losses. Farm yard manure and slurry should be brought out in quantities that the plants can take up in a short time. It should be worked into the topsoil soon after application.

Principle 2: Closed Nutrient Cycles

- Maximize recycling of plant residues, by-products, dung and farm wastes. Recycled or saved nutrients also mean money saved.
- Deep-rooting trees and shrubs planted in spare corners collect leached nutrients and can supply a great deal of mulch material if intense pruning is done.
- Compost can be made out of almost any organic material from the farm. It is not only a means of recycling nutrients but also increases the exchange capacity of the soil.
- Mulching is a simple way of recycling nutrients. It helps to keep moisture in the soil and feeds soil organisms.
- Ashes of stoves are a highly concentrated mixture of nutrients like potassium, calcium, and magnesium and may be applied to fields or mixed into the compost.
- Different plants have different requirements for nutrients; mixed cropping and crop rotations help to optimize the use of nutrients in the soil.

Principle 3: Optimize Inputs

- Introduce external organic 'wastes', if available. Several cheap organic wastes like coffee husks, sugarcane bagasse, rice husks, cotton stalks etc. may be available in the region and could be used to prepare compost.
- Chemicals like rock phosphate or dolomite help to supply scarce nutrients, and are less prone to leaching and less harmful to the soil than concentrates.
- Nitrogen fixing plants provide cost-free nitrogen. They can be planted as cover crops, food grains, hedges or trees, and also provide firewood, mulch and fodder.

3.2.3 Effects of burning plant materials

Burning is common in shifting cultivation and in the process of destroying agricultural wastes, as it saves labour. The ash contains nutrients, which are directly available to the plants. However, burning has many disadvantages:

- Large amounts of carbon, nitrogen and sulphur are released as gas and are therefore lost.
- The nutrients in the ash are easily washed out with the first rain.
- Plant materials are too valuable a source of soil organic matter to be burned.
- Burning harms beneficial insects and soil organisms.

In organic agriculture, plant materials shall only be burned as an exception (e.g. crops affected by diseases or hardy perennial weeds). Instead, they should be used for mulching or composting.

3.3 Fertility improvement

Farmers can improve the fertility of their soil by various management practices:

- Protection of the soil from strong sunlight and heavy rain by means of plant cover or mulch in order to prevent soil erosion and to preserve moisture.
- A balanced crop rotation or mixed cropping with a suitable sequence of annual crops grown on a field for preventing a depletion of the soil.
- An appropriate tillage method for obtaining a good soil structure without causing erosion and compaction.
- A good nutrient management by application of manures and permitted fertilizers according to the demands of the crops in their respective growth stages.
- Feeding and protection of soil organisms enhance the activity of beneficial soil microbes and organisms like earthworms by supplying organic material.
- To stabilize the structure, it is important to protect the soil surface with mulch or plant cover and to apply organic material (ideally compost).

Further Readings:

1. Conversion to Organic Agriculture by A.K. Singh. International Book Distributing Co., Lucknow.

Chapter 2

Composting

In organic farming, farmer has to manage the farm with coherent diversity by utilising all the on-farm and adjacent resources. There are vast potential of manurial resources and organic wastes available for recycling. Residual wastes from crops and natural biomass hold dependable promise for innovation in nutrient recycling in addition to their own optimum utilisation.

Compost is a balanced organic fertilizer and it provides both macro and micronutrients to plants. In addition, it also improves soil fertility, buffering capacity, porosity, aeration, temperature and water holding capacity. Compost is prepared from organic substances obtained from different sources like plant and animal wastes of the organic farm. Composted organic manures provide all the nutrients that are required by plants but in limited quantities. It helps in maintaining C:N ratio in the soil and improves the chemical, physical and biological properties of the soil. Due to increase in the biological activities, the nutrients that are in the lower depths are made available to the plants.

The principle of composting is based on decomposition of nature's life cycle and is facilitated by microorganisms. Composting converts organic substrates into utilizable forms in absence or presence of air to yield a sanitary soil supplement. The process can handle biodegradable crop wastes such as animal and vegetable wastes, leaves, aquatic weeds, chopped corn or rice/wheat stalks etc. with cowdung and soil.

1. Process of Composting

1.1 Selecting the primary materials

The composition of the composting material is of major

importance. The C/N-ratio and the structure of the material have a major influence on the composting process. Material that is rich in nitrogen (low C/N-ratio) does not usually contribute to a good structure and thus does not allow for good aeration if composted separately. Material which has a good structure usually has low nitrogen content (high C/N-ratio) and does not offer enough nitrogen for the bacteria to feed on. The mixture of different materials should contain an average C/N-ratio of 30:1. Smaller material must be mixed with more bulky material. To allow an ideal composting process, the mixture should consist of approximately:

- One third bulky material with a rich structure (chopped branches and tree bark, bulky material separated from previous composts)
- One-third medium to fine material with a high C/N-ratio (straw, crop residues etc.)
- One-third fine material with a low C/N-ratio (fresh leguminous, manure etc.), and
- About 5 to 10 % soil.

Material suitable for composting:

- Plant material: a balanced mixture of N-rich and C-rich material.
- Animal dung: cow, pig (rich in K and P), poultry (very rich in P), goat, horse etc.
- Wood ash: contains K, Na, Ca, Mg etc.
- Rock phosphate: the phosphorus binds to the organic material and is thus less fixed to soil minerals; it is therefore better applied to the compost heap than directly to the soil.
- Small quantities of soil, especially soil rich in clay, or ground rock improve the composting process and the quality of the compost. They are mixed with the other material or used to cover the heap to reduce nutrient losses.

Material NOT suitable for composting:

- Plant material affected by diseases like rust or virus.
- Persistent perennial weeds unless first dried in the sun.
- Material with hard prickles or thorns.
- Waste such as metal or plastic.

1.2 Setting up a compost heap

- Prepare the composting material properly. Chop coarse woody material to increase its surface area and encourage decomposition by fungi and bacteria.
- If dry, soak the composting material before mixing it.
- At the bottom of the heap, put twigs and branches to allow for good drainage of excess water.
- Pile up coarse carbon rich and nitrogen rich material in alternating layers.
- Manure or old compost applied to each layer enhances the composting process.
- Thin earth layers between the compost help to prevent nitrogen loss.
- A 10 cm thick cover of straw or leaves in the initial stage, and an impermeable cover (sacks, plastic sheet etc.) in the final stage prevent potassium and nitrogen being washed out of the heap. In dry climates, cover the heap with a 15 cm thick layer of mud.
- If the heap is not moist enough, from time to time pour water or liquid manure over the compost.

Two to three weeks after building up the compost heap, it used to decrease to about half its original size. This is the right time to turn it. Turning has a number of advantages:

- Turning the compost helps to accelerate the process.
- It improves aeration and encourages the process of composting.
- It ensures that material from the outside of the heap can decompose properly when placed in the center.

- It allows the quality of the composting process to be checked and for any non-ideal conditions to be improved.

2. Systems of Composting

There are several improved methods of compost making, which increase the rate of decomposition and minimize the losses of nutrients. The methods of composting have been researched both under aerobic and anaerobic conditions.

Anaerobic composting: In this system, the organic waste is decomposed in the absence of air. Organic waste may be collected in pits and covered with a thick layer of soil and left as it is undisturbed for 6-8 months. The compost so formed is of poor quality, as there is incomplete decomposition of organic waste, and it may include aggregated masses.

Aerobic composting: It is a process by which organic waste is converted to compost fertilizer in the presence of air. It can be done by different methods. In these methods proper movement of air through the organic waste have to be facilitated by turning and raking.

2.1 Indore method

This is the first aerobic method of composting in which manure is ready within four months. In this method animal dung is used (as catalytic agent) along with all other plant residues and waste materials available on the farm.

2.1.1 Materials for composting

The materials that can be used are:

- Mixed crop residues, plant leaves, weeds, grasses, wood ashs, bran, etc.
- Dung of all the farm animals like cows, buffaloes, bullocks, goats etc. collected with beddings.
- Urine soaked mud of the cattle shed
- Wood ashes
- Water and air

Hard woody materials are chopped in to small pieces and are cursed. Animal dung of all the farm animals like cow, buffaloes, bullocks, horses, goats etc. is collected with bedding. Wood ashes reduce acidity of the compost and add potassium. Water and air are necessary for bacterial and fungal activity.

2.1.2 Size of the composting pit

The size of the composting pit should have a width of 2 to 2.5 m and depth 0.9 m. Depending on the availability of waste materials, the length may be from 3 to 10 m. Composting pit plan of Indore method is depicted below:

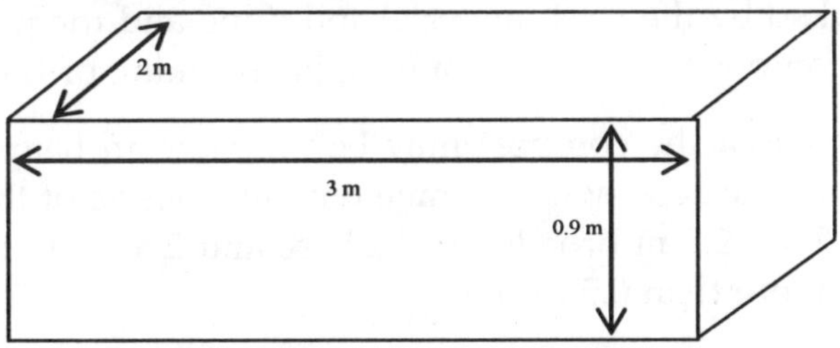

COMPOSTING PIT PLAN OF INDORE METHOD

The side of the pit should be sloppy and should have sufficient space between them for the carts to move freely.

2.1.3 Ingredient and methods of filling the pits

The materials are filled in the pit layer by layer.

- The first layer is of cattle shed waste filled to height of 7 to 8 cms with the help of racks. Wood ash (if available) will also be spread over, along with urine and mud.
- After this, 5 cm layer of bedding with cattle dung and soil are spread evenly. Sufficient amount of water is sprinkled over the material to make it moist.
- These layers are repeated to a height of 30 cm. It should not take more than six or seven days to fill the 3/4 length of the pit, leaving about one-fourth of the pit empty to facilitate subsequent turnings. The layer of bedding material is placed alongwith slurry and water.

- At the end one more layer of bedding material with wood ash and urinated mud should be provided.
- The materials are turned up three times.

  First turning: 10-15 days after filling the pit

  Second turning: 15 days after first turning

  Third turning: After 2 months
- At the time of first, second and third turning, the heaps are moistened with water. To maintain moisture to a level of 40 to 50 per cent, it is necessary to sprinkle water often in the evening and morning. In this way sufficient water is soaked by the trash material and dung and the process of decomposition starts, resulting in shrinking of the heap.

In rainy season the compost may be prepared in heaps above ground. In the areas with average rainfall, the size of the heaps may be 2.5 x 2.5 m broader at the base and 2 x 2 m at the top and not more than 0.5 m high.

In regions with heavy rainfall, compost should be prepared at safer places and protected by shed. Compost making should be avoided between the months of June and September.

This Indore method of composting was developed by A. Howard and Y.D. Wad at the institute of Plant Industry, Indore between 1924 and 1931.

2.2 Banglore method

This method overcomes many of the disadvantages of Indore Method. This method eliminates the necessity of turning up of the materials.

2.2.1 Ingredient and method of filling the pit

- To begin with, farm refuses are collected and spread in a 15 cm layer at the bottom of the pit; then it is moistened with water.
- Thereafter, a two inch layer of cattle dung and urinated mud is followed by a layer if 2-5 cm inch thick layer of soil.

- It is advisable to mix earth layer a suitable quantity of oil cake and bone meal.
- The process is repeated till the heap is 45-60 cm.
- Ultimately the heap is covered with mud plates about 2.5 cm thick.
- Compost will be ready within 8 to 9 months.

2.2.2 Size of the composting pit

Composting is done in trenches of 9 x 1.8 x 0.9 m or in pits of 6 x 1.8 x 0.9 m leaving 1.5m vacant spaces between successive trenches. The composting plan of Banglore method is depicted below:

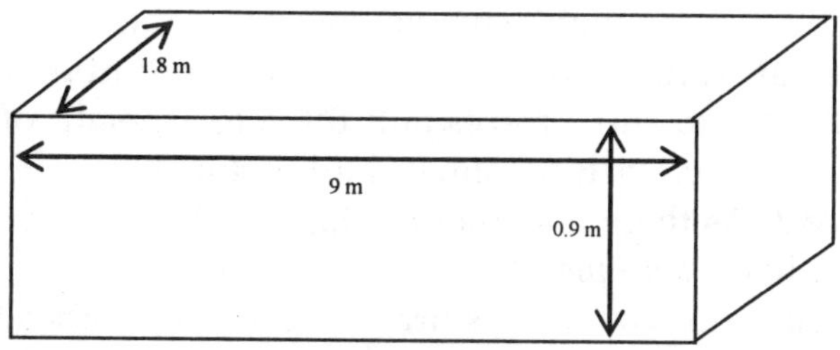

COMPOSTING PIT PLAN OF BANGLORE METHOD

2.3 VAT Method

VAT method is one of the improved methods of composting.

2.3.1 Size of the composting tanks

In this method, tanks are constructed using stones slabs and the dimensions of the tanks is 10 x 2 x 1 m. Open spaces of 2 to 4 cm are left between two slabs.

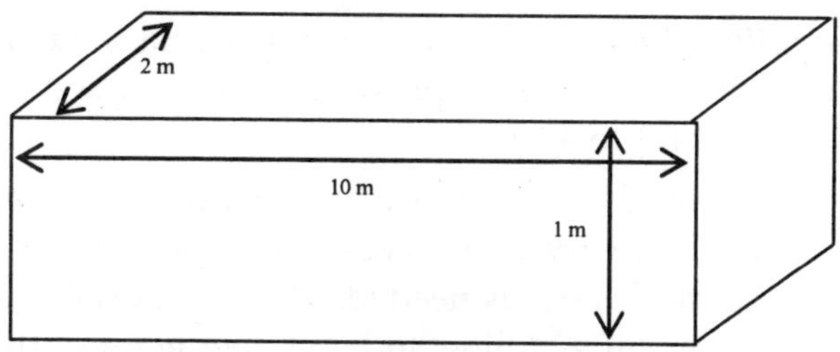

COMPOSTING TANK PLAN OF VAT METHOD

If stones are not available, locally available materials like bamboo poles, casurina poles or areca trees with coconut fronds can be used for construction of pits. But this type of pits will be temporary in nature. The floors of the tank are cemented to prevent infiltration of nutrients in to the soil.

The site selected for the tank should be easily approachable and at a comparatively higher level so that neither rainwater gets into it nor the water table rises and causes water stagnation in the tank during monsoon. It should be near the cattle shed and the source of water supply.

2.3.2 Ingredient and method of filling the tank

- The tanks are filled in layers.
- The *first layer* should have slowly degradable materials like coconut fronds, sugarcane thrash, coconut fibre or maize stalks cut to small pieces and filled to a height of 15 cm. Over this cow dung slurry and a thin layer of soil are spread. As these materials are highly fibrous, microorganisms like *Pleurotus, Aspergillus,* or *Trichoderma* should be mixed with cow dung slurry for early decomposition.
- The *second layer* should be of dry grasses, weeds and crop residues and filled to a height of 25 to 30 cm. After these materials are filled, water in sufficient quantity is added to make these materials wet. Over this layer, cow dung slurry with microorganism is applied as explained earlier.
- A *third layer* of green leaves to a height of 25 to 30 cm and a *fourth layer* of tank silt, bio gas slurry and other animal wastes like poultry manure, sericulture waste etc. are added to a height of 25 to 30 cm.
- The *fifth and final layer* of cow dung, cattle shed waste and animal urine is then applied and covered with soil to a thickness of 3 to 4 m.
- In this method, maintenance of moisture to a level of 50 to 60 per cent is very much necessary, and for this frequent application of water is needed. These materials are turned over 45 days after filling and covered again with 50 to 60

per cent of moisture level.

- Compost used to ready in about 90 to 100 days. In composting, green leaves from trees like pongamia, neem and weeds like water hyacinth, ipomoea, calotropis etc. can be used. Well-decomposed compost will have a stable temperature with no bad odour. The colour of the well decomposed compost should be brownish black. The pH will be around 7 and C:N ratio should be less than 15.

2.4 NADEP method

This method envisages lot of composting through minimum use of cattle dung. Decomposition process flows through 'aerobic' method and it requires about 90 to 120 days for obtaining the decomposed compost.

2.4.1 Size of the composting tanks

The method requires construction of a tank admeasuring 3 x 1.8 m or 3.6 x 1.5 m internally with 25 cm thick perforated brick wall all around in mud or cement mortar to a height of 0.9 m above ground.

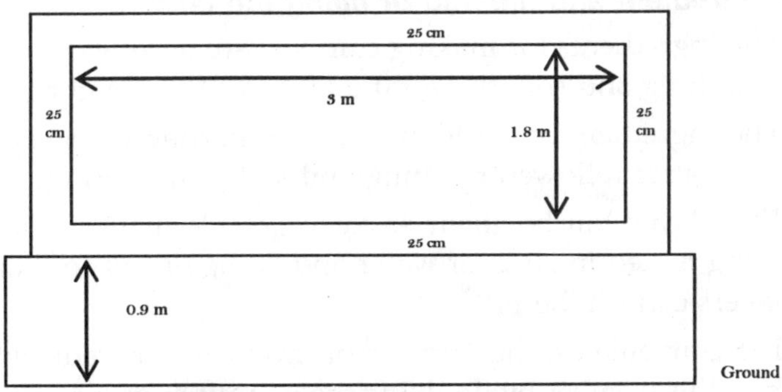

COMPOSTING TANK PLAN OF NADEP METHOD

The above ground-perforated structure facilitates passage of air for aerobic decomposition. The floor of the pit is laid with bricks. The tank is covered above with a thatched roof. This prevents loss of nutrients by seepage or evaporation and the contents are not exposed to sun and rain.

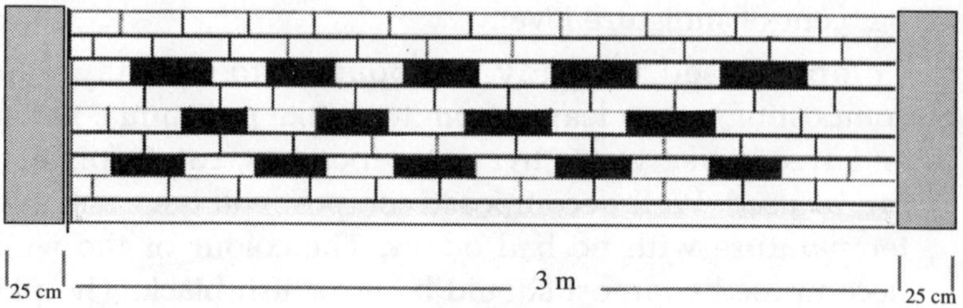

SECTION A-A

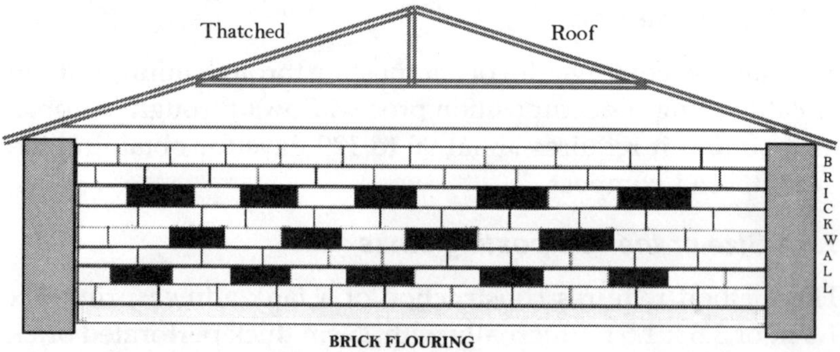

SECTION B-B

2.4.2 Ingredient and method of filling the tank

- The ingredients for making compost are agro-wastes, animal dung and soil in the ratio of 45:5:50 by weight.
- The ingredients are added in layers starting with vegetable matter followed by dung and soil in that order.
- Each layer can be about 45 kg vegetable matter, 5 kg of dung mixed in 70 lit of water and 50 kg of soil so that 30 layers can fill the pit.
- For convenience the number of layers can be reduced to half this number by doubling the quantities of ingredients in each layer.
- Tree loppings and green manure crops can also be used to fill up the tank if sufficient farm wastes are not available at time.
- The nutrients produced in the manure are absorbed by the soil layers thus preventing their loss.

- About 20-50 1it of water is to be sprinkled twice a week after the tank is loaded. The material loaded has to be left in the tank for about 100 to 120 days for complete decomposition of the material.
- One tank can be used three times a year. With production of 3 tons to 3.5 tons of compost produced per cycle, about 9 to 10 tons of compost can be made annually from one tank.
- The compost can be stored for future use, preferably in a thatched shed after air drying and maintaining it at about 20% moisture level by sprinkling water whenever needed.
- It is advisable to sprinkle cultures like *Trichoderma, Azatobacter* and PSB in layers to enhance the speed of composting process. The microorganism present in the material will utilise the oxygen in the air and convert the organic matter into compost.

By following the procedures suggested above, the compost could be preserved for about 6 to 8 months. It is very simple to construct and easier to operate. In this method, compost can be prepared with minimum quantity of cow dung use and hence, it can be considered as very versatile model.

NADEP method of composting was developed by Shri N.D. Pandhari Pande from Maharashtra. The compost made out of this process has been tested by several institutions like IIT, New Delhi, Gandhigram University, Centre for Science, Wardha etc. including the farmer's field. This method takes care of all the disadvantages of heaping of farm residues and cattle shed wastes, etc. in the open.

2.4.3 Unit size and cost

The following unit costs are suggested for construction and operation of composting tank per unit. It is necessary that a farmer should have at least 2 tanks so that when one is filled up the other one is available for loading the material generated in his farm. Thus, for a farm size of 5 acres dryland a unit of two tanks is needed. If the farmer is having mixed farm of dry land and irrigated farm, one should have 4 tanks.

Construction cost

Particulars	Cost in Rs.
Bricks 1200 nos @ Rs.1500/1000	1800
Cement 200 kgs @ Rs.5.00 per kg	1000
Sand 3 m^3 @ Rs.150/ per m^3	450
Masons 3 No. @ 150/ per day	450
Labourers 8 No. @ Rs. 100/ per day	800
Light thatched roof, LS	700
Total	5200

Operational cost

Particulars	Cost in Rs.
Cow dung 200 kg @ Rs.0.50/kg	100
Agro-waste 1350 kg @ Rs.0.25/kg	335
Cost of digging soil and transportation	300
Water sprinkling charges - Once in 4 days	300
Cost of tank filling 2 labours half day	100
Cost of unloading, removing undecomposed material etc.	100
Miscellaneous	100
Maintenance cost of Tank per year	500
Total	1835

2.5 Water hyacinth compost

Water hyacinth (*Eichhornia crassipes*) is an aquatic weed and grows luxuriantly in ponds, lakes, and water reservoirs. It can be used as soil mulch, green manure and compost in agriculture. Water hyacinth can be converted in to manure without any additional cost.

2.5.1 Size of the pit

Pit Size: 2 x 1.5 x 0.5 m

2.5.2 Method of filling the pit

- Water hyacinth are collected from water bodies and are

spread to the ground for a day or two to reduce the water content of it.

- First, a layer of water hyacinth is spread at the bottom of the trench and over this cow dung mixed water is sprinkled to hasten decomposition.
- The process is repeated till the pit is filled to 50-60 cm high above ground level. The top of the heap should be made dome shaped and plastered with 2.5-50 cm thick layer of soil mixed with cowdung.
- Compost will be ready in about 2 months.

The fresh water hyacinth weed contains about 3.5 per cent organic matter, 0.04 per cent nitrogen, 0.06 per cent phosphorus and 0.02 per cent potash. In comparison to Farmyard manure water hyacinth compost is richer in potash.

2.6 Compost Tea

Compost tea or liquid manure can be prepared by decomposing the green leaves of legumes.

2.6.1 Ingredients

- Green leaves of legumes: 5 to 7.5 kg
- Cow dung or droppings of goat, sheep or poultry: 5 to 7.5 kg.
- Drum: One
- Gunny bag: One
- Water: 200 lit

2.6.2 Method of preparation

- The cow dung or droppings of goat, sheep or poultry are mixed in the green leaves of legumes.
- These materials should be kept in the gunny bag which is kept in the drum having 200 lit of water.
- A brick is placed on the gunny bag to keep it on the bottom of drum.

- The water of the drum should be stirred for a period of five minutes every day and this is to be continued up to 20 days.
- The manures become ready within three weeks.
- When the liquid manure is fully prepared, no bad odour should be smelt out from the drum and colour of water of the drum should be as like of tea.
- Now, gunny bag is taken out from the drum.
- 200-250 gms of molasses (gur) is added to the mixture to increase the quality of the liquid manure.
- Equal amount of water should be added with the mixture before use.
- The quality of manure remains good for a period of 4-6 weeks, if it is preserved in cool and shady place.

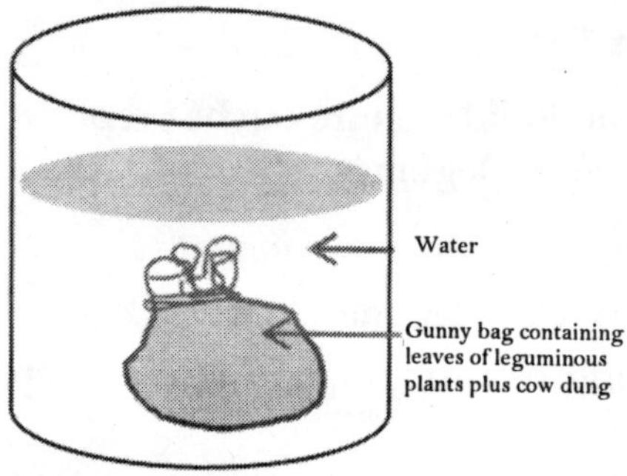

2.6.3 Use of compost tea

- The liquid manure can be applied to the crop in the morning or in the late afternoon with drip irrigation water.
- It can also be used as foliar spray.

Compost when added to soil effect the soil physically, chemically and biologically and bring a number of beneficial actions on soils. Moreover, compost is rich in microorganism specially bacteria and fungi and when incorporated in soil, not only millions of organism present in it are added but it activates other already present in soil. The water holding and heat absorbing capacity of soils also increases and thus improves the permeability of soils.

Chapter 3

Vermicomposting

Vermi-composting provides the nutrients and growth enhancing hormones necessary for plant growth. The fruits, flowers and vegetables and other plant products grown using vermi-compost are reported to have better keeping quality. Vermicomposting involves application of earthworms for recycling of organic wastes and their conversion into nutrient rich organic fertilizer. Earthworms eat soil and various kinds of organic matter which undergo complex biochemical changes in the intestine, which are then excreted out in the form of granular casts known as vermicastings. The vermi-casting containing nutrients are rich manure for the plants. The organic carbon in vermicompost releases the nutrients slowly and steadily into the system and enables the plants to absorb the nutrients.

1. About the Worms

Of about 350 species of earthworms in India with various food and burrowing habits, *Eisenia fetida, Eudrilus eugeniae, Perionyx excavatius* are some of the species for rearing to convert organic wastes into manure. Commonly known earthworms employed for vermin-casting are given below:

Ecoform	Species	Origin
Epigeic	*Eisenia fetida* *Eudrilus euginiae* *Perionyx excavatius*	Exotic
Anecic	Lampito mauritti	Native
Endogeic	*Metaphire posthuma* *Octochaetona thurstopni*	Native

Basic characteristics of suitable species: Worm should be efficient converters of plant or animal biomass to body proteins, so that its growth rates are high. Earthworm should have high consumption, digestion and assimilation rates. It should be wide adaptability to environmental factors and have feeding preference for wide range of organic material. Worm should produce large number of cocoons that should not have long hatching time, so that multiplication and organic matter conversion is fast. Growth rate and maturity from young one to adult should also be fast. Earthworm should be disease resistant and have compatibility with other worms at different strata.

Multiplication of earthworm: One earthworm reaching reproductive age of about six weeks lays one egg capsule (containing 7 embryos) every 7 - 10 days. Three to seven worms emerge out of each capsule. Thus, the multiplication of worms under optimum growth conditions is very fast. The worms live for about 2 years. Fully grown worms could be separated and dried in an oven to make 'worm meal' which is a rich source of protein (70%) for use in animal feed.

Composting of organic waste using three species of earthworms

Earthworm species	Organic waste (kg)	Composting Time (days)	Compost recovery (kg)	Population increase of young ones
E. eugeniae	200	60	160	32,000
E. fetida	200	90	160	30,000
P. excavatus	200	90	160	6,000

Source: Kale and Sunita, 1993

2. Requirements for Vermicomposting

Following are the items required to be considered while setting up a unit for production of vermi-compost.

Location: Vermicomposting units should be ideally suited to locations with generation of considerable quantities of organic wastes. Suburbs of cities and villages around urban centres can be ideal locations for practice of vermicomposting on a large scale, from the viewpoint of availability of raw material and marketing of the produce. As use of the compost is said to have ameliorative effect on product from fruit, flower and vegetable crops, vermicomposting units may be located in areas with concentration of fruit and vegetable growers and floriculture units.

Land: About 0.5-1.0 acre of land will be needed to set up a vermiculture production-cum-extension centre. The centre should have at least 8-10 sheds each of about 180-200 sq.ft. It should also have a bore well, and pump set or watering arrangement and other equipments.

Sheds: For a vermicomposting unit, whether small or big, this is an essential item and is required for having the vermi beds. They could be of thatched roof supported by bamboo rafters and purlins, wooden trusses and stone pillars. If the size is so chosen as to prevent wetting of beds due to rain on a windy day, they could be open sheds. While designing the sheds, adequate room has to be left around the beds for easy movement of the labour for filling and harvesting the beds.

Vermi-beds: Normally the beds are 75 cm - 90 cm thick depending on the provision of filter for drainage of excess water. The entire bed area could be above the ground. Care should be taken to make the bed with uniform height over the entire width to the extent possible to avoid low production owing to low bed volumes. The bed width should not be more that 1.5 m to allow easy access to the centre of the bed.

Seed stock: The worms multiply fast to give the required numbers over a period of 6 months to a year. Thus, worms @ 350 worms per m^3 of bed space should be adequate to start with and to build up the required population in about two cycles or three.

Water supply system: As the beds have always to be kept moist with about 50% moisture content, there is need to plan for a water source, lifting mechanism and a system of conveying and applying the water to the vermi-beds. Drippers with round the clock flow arrangement would be quite handy for continuous supply and saving on water. Such a water supply/application system requiring considerable initial investment, however, reduces the operational costs on hand watering and proves economical in the long run.

Machinery: Farm machinery and implements are required for cutting (shredding) the raw material in small pieces, conveying shredded raw material to the vermi-sheds, loading, unloading, collection of compost, loosening of beds for aeration, shifting of the compost before packing and for air drying of the compost, automatic packing and stitching for efficient running of the unit.

Buildings: When the activity is taken up on a large scale on commercial lines, a building is required to house the office, store the raw material and finished product.

Transport: For any vermicomposting unit transport arrangement is a must. When the source of raw material is away from the production unit, an off-site transport becomes major item of investment. A large sized unit with about 1000 tonnes per annum capacity may require a 3-tonne capacity mini-truck. With small units particularly with the availability of raw material near the site, expending on transport facility may become infructuous. On-site transport facilities like manually drawn trolleys to convey raw material and finished products between the storage point and the vermicompost sheds is also required.

Furniture: A reasonable amount could also be considered for furnishing the office-cum-stores including the storage racks and other office equipment. These enhance the efficiency of operations.

Fencing and Roads/Paths: The site area needs development for construction of structures and development of roads and pathways for easy movement of hand-drawn trolleys/wheel barrows for conveying the raw material and the finished products to and from the vermi-sheds. The entire area has to be fenced to prevent trespass by animals and other unwanted elements.

3. Production of Vermicompost

Vermicomposting process involves three steps- pretreatment, composting and residue disposal. Pretreatment involves receiving refuse, shredding, decreasing its particle size and volume, separation, storage and transfer.

Vermicomposting materials: The worms feed on any biodegradable matter ranging from coir waste to kitchen garbage. All biologically degradable and decomposable organic wastes are used in vermicomposting. These include animal dung, agricultural wastes, forestry wastes, leaf litter, waste paper, cotton clothes, city garbage, bio-gas slurry and industrial wastes of biological nature. Materials with high nitrogen or proteins are best activators. High carbohydrates, if present, can lead to acidification. Ideal pH is 6.5-7.5.

Pretreatment of composting materials: The materials collected for vermicomposting should be free from non-degradable materials like plastics, stones, glass, ceramics and metals. Agricultural wastes may be required to cut into smaller pieces for enhancing decomposition process.

The separated material should be spread in a layer of maximum of 1 foot to expose to sun for a day. To control insect pest infestation, if any, neem formulation can be used. This helps in killing some of the unwanted organisms and removes foul smell. Then the material should be mixed with cowdung slurry and cellulolytic inoculants, made a heap covered with hesian clothes or left on ground for 4/5 weeks for partial decomposition. This partially decomposed materials are now ready for vermicomposting.

3.1 Method of vermicomposting

Pit/Tank Size: The earthworms being voracious eaters consume the biodegradable matter and give out a part of the matter as excreta or vermi-castings. A composting pit of any convenient size can be used for Vermicomposting.

Preferably the pit size may be as big as 15 m x 1.5 m x 0.3 m. Alternatively, for small farmer a pit of 1.5 m wide x 1 m height will give a low cost production unit. The pit should be preferably of concrete structure.

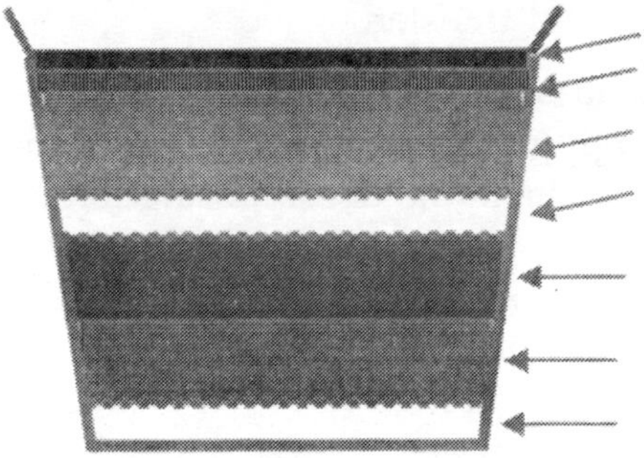

Cover of feed substrate: The pit/tank should be filled with feed mixture as follows:

- At the base of pit/tank, a layer of broken bricks is to be placed, followed by coarse sand. The thickness of layer should range between 5-7.5 cm well suited for drainage of excess water.
- At second layer, straw of paddy/wheat, banana stem peels, coconut leaves, sugarcane trash, crop stems, grass or husk should be placed. The thickness of layer should be about 30 cm well suited for aeration.
- Third layer should be of 15-30 days old cow dung with the thickness of about 20-30 cm which acts as reserve food for earthworms.
- Fourth layer or top layer should have partially decomposed waste up to thickness of 30-37.5 cm which is used for composting.
- Earthworms are introduced in between the layers @ 350 worms per m^3 of bed volume. Pit is then covered with moist gunny bag. Moist gunny bag used for reducing the moisture loss and also save worms from predators like ants.
- The beds are maintained at about 40-50% moisture content and a temperature of 20-30° C by sprinkling water

over the beds. If moisture is high, dry cowdung or leaf litter should be mixed in the substrate. The pH of the substrate should be between 6.8-7.5.

- Sprinkling of water should be stopped before 3-4 days of harvesting to allow the worms to go down because of the drying of surface layers and the compost is then harvested, dried in shade and packed. The collected worms can be released in freshly prepared beds.
- The bed should be under roof to prevent direct sun and rain.

3.2 Vermicast Collection

The earthworms fragment the waste, thereby increasing the surface area and encouraging further microbial activity. The earthworm derives their food from microorganisms in the waste and from the organic matter that is being decomposed. The methods for collection of vermicastings are as follows:

- The earthworm that feed actively assimilate only 5-10 percent of the feed and rest is excreted as loose granular mound of vermicasting on the surface, generally away from the food source. These vermicastings have to be brushed aside and collected into separate trays.
- The collected vermicastings have to be left overnight in conical heaps for the worms to move to the bottom. The tops of the cones which are free from worms are then collected and lightly air dried.
- The dried vermicastings are sieved through 3 mm mesh to separate cocoons and young ones from the vermicastings. The dried and sieved vermicastings are now ready for use as manure.

One earthworm eats approximately 1 g/day and same quantity of faeces is released. Thus 350 worms will produce 350g compost per m^3 of bed volume per day. During the process of vermicomposting the population of earthworms can increase by 20-25 times.

3.3 Vermiculture

Vermiculture is the process of rearing and breeding earthworms. The type of earthworm species, soil condition, food preference and behavioural response of earthworm counts the success of vermiculture technology.

Earthworm release its cocoons in the soil. The cocoons then incubate in about 18 days into juveniles. The juveniles are the larval stage of earthworm. The juveniles then transformed into non-clitellate state. A clitellum is developed on worm at the last growth phase. After this growth stage, larvae go into reproductive stage under which the worms copulate. Later, the worms begin shedding their cocoons in soil in about 10 days. This is a general life cycle pattern of earthworms.

Earthworm production can be done in pot by releasing a large number of earthworms on fully decomposed organic materials. Important steps adopted under pot methods are given below:

- Take a china pot of size 30x45 cm and filled it with 10 kg of fully decomposed organic material or FYM. The organic matter should contain proteins cellulose and at least 1% nitrogen, (e.g. grass eater cattle dung).
- Release 1000 adult earthworms in the pot and cover it with moist gunny bag.
- After 7-10 days remove all the feed with earthworms from the pot. Separate out the earthworm.
- Fill the pot by fresh organic material (fully decomposed) and release the same worms.
- At every removal, on an average 500-600 cocoons will be obtained. Average length and width of cocoons is 6-8 mm, more or less oval in shape and tapers at the ends. They are initially white in colour then turn light green, pale brown and dark brown or reddish brown before hatching.
- These cocoons are further reared separately for juveniles (larvae) and adult earthworms in sufficient quantity of organic materials.

- Give the food to the earthworms as per their need. About 1 kg organic material is sufficient to the worms weighing about 2000 flat worms and 4000 reproductive worms.
- Maintain proper moisture by covering the pots with moist gunny bag. White colour of earthworms indicates excess of water and acidity in the pot.
- Flies and other insects disturb vermiculture. Hence, cover the pot with gunny bag. Neem seed extract may be mixed superficially on the surface of organic material to control the insects.
- Sick worms should be collected and disposed. A good quality culture has a typical smell but not a bad smell.

3.4 Vermiwash

Vermiwash is a liquid fertilizer collected after passage of water through a column of worm activation. It can be used as foliar spray after diluting with water. It can also be used as pesticides if mixed with 10% diluted urine of cow.

Preparation of vermiwash: Take an empty bucket with bucket cover and make a hole of 1 inch diameter in the side of the bucket at the bottom. Attach a tap in the hole with the help of T-tube joint as shown in figure. Keep the tap open till the bucket is filled up with materials required for preparation of vermiwash.

At the bottom of the bucket, place 20-25 cm thick layer of broken bricks or pebbles. Then arrange another 25-30 cm layer of coarse sand on the bottom of layer. A layer of 35-40 cm of loamy soil is then placed. This layer is kept moisten. Now release 100-120 earthworms and spread cattle dung over this layer of earthworms. Placed hey or partially decomposed plant residues on the top of the layer. Moisten this layer gently and the tap is closed after removal of excess water from the unit. The unit has to be moisten everyday. The top of the bucket should be kept open for about 30 minutes everyday. Unit will be ready after 15 days and then tap is closed.

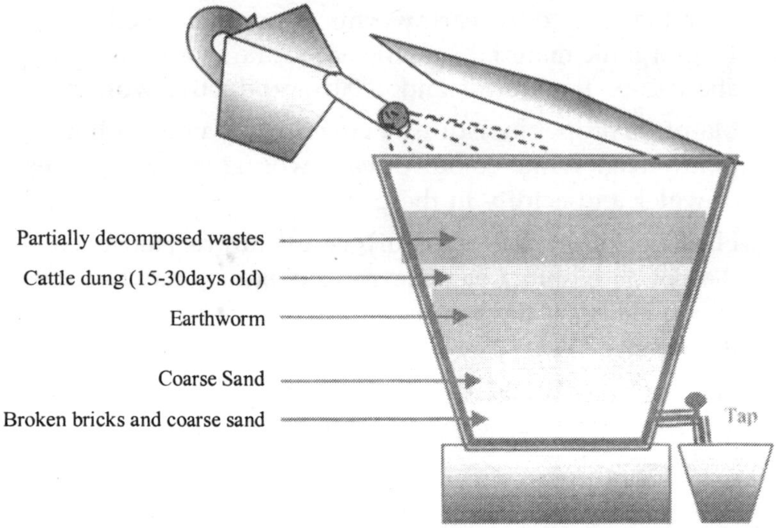

Vermiwash Unit

From next day, water should be sprinkle with the help of five liter metal water container having numerous perforated holes (generally used for watering to flower plants). Water will percolate through the compost and carry nutrients from freshly formed vermicasting through the filter unit. Vermiwash can be collected in a small container or bucket by opening the tap. After collection of vermiwash the tap is again closed and suspended bucket is refilled with water for collection of vermiwash of next day.

4. Costs of Production

Vermicomposting could be taken up on any scale starting from 10 tonnes per annum (TPA) to 1000 TPA and above. As the production is proportional to the vermi-bed space, it is advantageous to start with less capacities and later expands the unit after gaining production experience and developing assured market for the product.

A bed volume of 330 m^3 spread over sixteen beds (15 m x 1.5 m x 0.3 m) is estimated to produce vermicompost of 200 TPA over

5 cycles/crops of 75 days each annually. These beds are housed in 8 open sheds of 15m x 5.4m.

The requirement of materials and implements, machinery are listed below:

(a) Implements and machinery for 200 TPA vermicompost unit

Sl. No.	Particulars of item
1	Shovels, spades, crowbars, iron baskets, dung fork, buckets, bamboo baskets, trowel, wire mesh sieves (3 mm and 6 mm)
2	Plumbing and fitting tools
3	Power operated shredder
4	Sieving machine with 3 wire mesh sieves 0.6 m x 0.9 m size - power operated without motor
5	Weighing scale (100 kg capacity)
6	Weighing machine (platform type)
7	Bag closer
8	Empty barrels (200 L capacity) 4 Nos.
9	Culture trays (plastic) (35 cmx 45 cm) - 4 Nos.
10	Wheel barrows - 2 Nos.
11	Agricultural waster @ 320 kg per m^3 (105.6 ton capacity)

(b) Materials required for construction of temporary shed for setting up 200 TPA vermicompost unit (Size 8 m x 15m x 5.4 m)

Sl. No.	Particulars	Quantity
1	Wooden ballies (3 m long)	472
2	Wooden ballies (3.6 m long)	48
3	Bamboos (3 m long)	800
4	Bamboos (6 m long)	240
5	Bamboo mats for covering the roof	720
6	Coir rope 6 mm dia	200 kg
7	Binding wire for tying bamboos/mats	100 kg
8	Labour charges for erection of sheds	--
9	Miscellaneous	--

(c) Bedding materials requirements

Sl. No.	Particulars	Unit
1	Cow dung @ 80 kg/m^3	26.4 ton
2	Worms @ 350 per m^3 500 worms per kg	231 kg
3	Formation of vermin bed with agro-waste, cow dung and worms	330 m^3

(d) Costs and Benefits (200 TPA vermi-composting unit)

Sl/No.		Particulars	Years	
			Ist	IInd onwards
1		Costs		
a)		Investment costs		
	i)	Open sheds with bamboo mat roofing	72000	-
	ii)	Machinery and tools	80000	-
	iii)	Office-cum-Store	60000	-
	iv)	Water source	60000	-
	v)	2 NADEP tanks	10000	-
b)		Operational cost (For 5 cycles in a year @ Rs. 72,000 per cycle of 75 days)	360000	360000
		Total	650000	360000
2		Benefits		
	a)	Sale of vermicompost of 200 tonnes @ Rs.2500/- per ton (60% in first year and 90% from 2nd year onwards)	300000	450000
	b)	Sale of worms @ 5 kg per tonne of compost and Rs. 50 per kg	-	45000
		Total	300000	505000
		Net Benefit	--	137000

Production of 60% in the first year and 90% in the subsequent years is assumed. Benefits include the income from sale of vermicompost @ Rs.2500 per tonne and worm @ Rs.50/- per kg. The net income from the 2nd year onwards would be about Rs.1,37,000 annually.

When the commercial scale production is aimed at in addition to the cost of production, considerable amount has to be invested initially on capital items. The capital cost may work out to about Rs.1500 to Rs.2500 for every tonne of compost produced annually. The high variability in the unit capital cost is due to the fact that large units require considerable expenditure on machinery and transport particularly when the source of raw materials is away from the site of production facility and the finished product has to be transported to far off places before being marketed.

The Society for Preservation of Environment and Quality of Life (SPEQL) organised a pilot project on vermicomposting in fruit market premises, Kothapet, Hyderabad. The project, being operated currently on commercial lines is serving as a demonstration unit. The estimates of costs and benefits, presented here are based on the experience of that pilot project.

Income from Extension Service: The unit will provide extension services to the near by villages. Under this the unit will provide cultural material of the desired species, and train farmers and entrepreneurs who would like to set up their own small units for use in their organic farms. Those who want to set up commercial units also can get know-how and culture material at a reasonable cost. The following benefits can be assumed under extension services for the unit:

- Sale of culture material @5-10 paise
- Consultancy for setting up new units @Rs.1000/- per unit and say 10 units per year comes to Rs.10000/-.
- It is advised that the some simple method for vermicomposting for example, NADEP compost process to serve as demonstration to the local farmers. Two NADEP tanks of size 10x6x3 feet are constructed at a suitable location.

5. Advantage of Vermicompost on Organic Farm

Vermicompost provide a natural organic soil amendment. They are rich in microorganism especially bacteria and free growth regulators. Moreover, the heavy metals in the waste get accumulated in the earthworms, the fertilizer so obtained comparatively gets rid of the heavy metals. It has been reported that earthworm, *Eisenia foetida* can accumulate about 150 ppm to 300 ppm of cadmium, nickel, lead and zinc.

Nutrient content increase and reduction in lignin in the vermicompost decomposed by earthworm *Perionyx excavatus.*

Organic wastes	No. of times increase			Times reduction in lignin
	Nitrogen	Available Phosphorus	Available Potassium	
Kitchen waste	2.38	9.92	1.37	13.33
Paddy straw	2.60	14.88	2.06	14.36
Maize straw	3.43	14.44	1.61	14.11
Banana leaves	3.65	11.42	1.18	13.33
Maize cobs	5.83	18.33	2.26	8.60
Titachampa litter	4.06	9.83	1.65	7.72
Moss	2.50	7.50	2.05	10.10
Pine needle	2.50	4.00	1.54	4.44
Coconut coir	4.07	6.00	1.54	6.50

Source: Kale 1998

Vermicompost, apart from supplying nutrients and growth enhancing hormones to plants, improves the soil structure leading to increase in water and nutrient holding capacities of soil. Earthworms convert nitrogen to ammoniacal nitrogen or nitrate nitrogen, phosphorus, potash and magnesium into available forms. As the waste is broken down, its particle size decreases progressively and its moisture holding capacity increases and becomes equivalent to that of peat. The surface area of the material increases which in turn helps as base for nutrients. The C/N ratio also deceases favouring plant growth and increasing value of waste.

Chapter **5**

Biofertilizer Production

In view of the priority for the promotion of organic farming and reduction of chemical residues in the environment, special focus has to be given for the production of biofertilizers. Biofertilizers enhance the nutrient availability to crop plants by processes like fixing atmosphere N or dissolving P present in the soil and also impart better health to plants and soil thereby enhancing crop yields in a moderate way. It is a natural method without any problems like salinity and alkalinity, soil erosion etc.

Biofertilizers, in strict sense, are not fertilizers which directly give nutrition to crop plants. Biofertilizers are carrier based preparations containing mainly effective strains of some microorganisms like bacteria and fungi when incorporated with seed or soil are capable of fixing atmospheric nitrogen or solubilising insoluble phosphate in soil and making them available to the crop plants. Their mode of action differs and can be applied alone or in combination. By systematic research, efficient strains are identified to suit to given soil and climatic conditions. They are packed in carrier materials like peat, lignite powder etc. in such a way that they will have sufficient shelf life.

1. Biofertilizer and their Beneficiary Crops

Biofertilizers have been broadly divided, based on their nature of functions like *Rhizobium, Azotobacter, Azospirillum,* etc. which have the ability to fix atmospheric nitrogen through the nitrogenase enzyme, which in turn enhance the nitrogen nutrition to the beneficiary crop are used under the name of Nitrogenous biofertilizers. On the other hand, bacterial cultures of *Bacillus, Pseudomonas* etc. as well as fungal cultures like *Penecillium* and *Aspergillus* solubilize insoluble phosphates present in the soil into soluble forms through secretion of organic acids and thus, improve the phosphate availability to the crops and are used as Phosphatic biofertilizers. However, the biofertilizers are mostly crop specific which are dependent on the compatibility of the microorganisms with the beneficiary crop for delivering maximum utility.

Biofertilizers and beneficiary crops

Name of the Biofertilizer	Benefits	Beneficiary Crops
Rhizobium	10-35% yield increase, Adds 50-200 kg N/ha.	Legume crops like Groundnut, Soybean, Lentil, Red-gram, Green-gram, Black-gram, Cowpea, Bengal-gram and Fodder legumes
Azotobacter	10-15% yield increase, Adds 20-25 kg N/ha	Vegetables, Rice, Wheat, Mustard, Safflower, Niger, Sunflower, Barley, Ragi, Jowar, Fruit & Plantation Crops, Spices, Condiment, Cotton, Mulberry, Ornamental Flowers
Azospirillum	10-20% yield increase	Sugarcane, Vegetables, Maize, Pearl millet, Rice, Wheat, Fodders, Oil seeds, Fruits and Flower
Phosphate Solubilizer	5-30% yield increase	All crops

2. Biofertilzer Production Technology

Different microbial organisms occur in soil habitat, however, their number are often inadequate in nature to effect increased crop production by supplying of crop nutrients biologically. Hence, it necessitates inoculating, artificially cultured microorganisms in adequate number so as to fix atmospheric nitrogen biologically or solubilise/mobilizing soil phosphates and supply crop nutrition to the beneficiary crops.

The agriculturally important microorganisms are used in biofertilizer production which are either natural isolates obtained by selection process or strain characteristics improved by mutation or biotechnological protocols. The contribution of biofertilizer, primarily depends on the efficacy of microbial strain and hence, it is desirable to choose a promising bacteria for use in commercial inoculants by the biofertilizer producers with strain characteristics like:

- fix adequate nitrogen or solubilise phosphate with high ability for colonizing in the root rhizosphere, form nodule in case of *Rhizobium*
- ability to grow well in artificial media with genetic stability
- ability to remain viable in the inoculants carrier material
- ability to tolerate environmental and persist in soil stress
- compatibility with host or range of crops

The selected strains are mass multiplied under aseptic conditions in a suitable fermenter, so as to get the desired population of 10^8 or 10^9 and quality of the broth in terms of microbial load as well as freedom from contamination are checked and the broth mixed with pre-sterilized carrier material. The carrier material should have high water holding capacity, non-toxicity to the microorganisms, good bufferingcapacity, easily sterilizable and having particle size of 150-200 micron. The mixing of broth with carrier material and packing are done under aseptic conditions. The biofertilizers produced are subjected for quality assurance by observing series of microbial protocols to check the microbial inoculums load of 10^8 or 10^9 per gram of carrier material besides authentication of the strain using conventional or serological methods before the biofertilizer packets are marketed by the biofertilizer producers.

Scientific aspects of production are standardised by Agricultural Universities and Research Laboratories. Machineries and laboratory equipments are available from various manufacturers and are of BIS standards.

2.1 Quality standards of biofertilizers

In association with the experts of BIS authorities, National Biofertilizer Project prepared (2000-2002) quality standards of Rhizobium, Azotobacter, Azospirillum and Phosphorus Biofertilizers (PSB). All these standards have been notified by BIS as under:

Rhizobium: (IS: 8268-2001): Cell count 10^7/g carrier till six months from date of manufacture (DOM); No contamination at 10^5 dilution, pH 6.4-7.5, carrier mesh 150-212 micron.

Azotobacter: (IS: 9138-2002): Cell count 10^7/g carrier till six months from date of manufacture (DOM); No contamination at 10^5 dilution, pH 6.4-7.5, carrier mesh 150-212 micron.

Azospirillum: (IS: 14807-2000): Cell count 10^7/g carrier till six months from date of manufacture (DOM); No contamination at 10^5 dilution, presence of white pellicle, pH 6.4-7.5, carrier mesh 100 micron.

Phosphate-Solubiliser (IS: 14807-2000); Cell Count 10^7/g carrier till six months from date of manufacture (DOM); No Contamination at 10^5 dilution, solublisation zone minimum 10 mm, pH 6.5-7.5, carrier mesh 100 micron.

The biofertilizer producers are required to strictly follow these standards. Biofertlizer package bags should be marked with name of the product, name and address of the manufacturer, batch number, date of manufacture, date of expiry, crop(s) for which intended with storage and usage instructions.

2.2 Requirements of biofertilzer production

In line with the technology and objective of biofertilizer production, various facilities are required as indicated below:

Land: It is required to set up laboratory and other facilities and office. Space may also be required for installing tube well/dug well and parking of vehicles. A minimum of half acre of land is required for setting up a 150 TPA (tonne per year) unit. Preferably, the entire site should be fenced with gates at suitable places.

Layout and buildings: The civil works comprises of building for laboratory, carrier preparation and enrichment, sterilisation, inoculation and quality control, maturation of culture, mixing and packing, storage/staff etc. The total covered area of about 3000 sq ft is required for the product manufacturing and other utilities.

Plant and machinery: Manufacture of biofertilizers needs a good number of laboratory equipments as well as other production facilities such as fermentors, culture medium tank, fermentor assembly, autoclaves, boiler, broth dispensers for sterlisation, demineralizing plant, air compressor etc,. All the machinery is manufactured in the country.

Manufacturing process and source of technology: The mother culture of various strains of biofertilizer is supplied from Agricultural Universities and Regional Centres of Organic Farming. The unit generally comprises of media preparation room, media store room, inoculation room, growth room, culture transfer room, sterilization, mixing and packing, etc. The floor plan should be designed to promote maximum efficiency and minimum contamination. The design should facilitate maintenance of optimum temperature, humidity and ventilation. The unit should be free from dust particles.

Infrastructural facilities for raw material, carrier material and utilities: The raw material required for biofertilzer production include ingredients for growth medium for the production of broth, carrier, packing materials like polythene packets, corrugated boxes, etc.,

i) Power: Normally a three phase electric supply is required for these plants. The normal requirements of a 150 TPA unit is about

70 HP. Depending upon the position of power supply, stand by generator may be needed.

ii) Water: A Biofertilizer production unit requires water mainly for steam generation for sterilization of carrier, broth preparation and cleaning of equipments. Accordingly well/bore well of designed size and according to the quality of water demineralization equipments are to be installed. The average per day requirement of water for 150 TPA capacities will be about 2500 to 3000 liters.

iii) Compressed air: It will be required for various pneumatic operations as well as for controlled air supply to fermenters, sterlisation/cleaning operations etc.

iv) Vehicles: The vehicles are required for procurement of carrier material and distribution of biofertilizers as well as for office use.

v) Manpower: For a unit manufacturing 150 TPA biofertilizers, the requirements of manpower are- Microbiologist, Floor Supervisor, Technical Staffs (Boiler operation, Mechanical maintenance, Packing machine operations, Electrical maintenance), 2-3 skilled and 4-5 semi-skilled labourers depending upon the volume of production

Environmental aspects and pollution control: No hazardous effluents are generated from a biofertilizer unit.

3. Commercial Production of Biofertilizers

Layout of the production unit: The biofertilizer plant should be housed in a suitable building complex. The main production unit should have separate channels for bacteriological work, carrier making and mixing and customer/visitor/marketing way. In addition there should be rooms with separate entrance for utilities like power, steam generator and stores. Appropriate design can be adopted in consultation with scientists/engineers.

Raw material: The chief raw materials needed for the production of biofertilizers are-

- Mother cultures
- Carrier material - lignite or bentonite or peat of desired quality in powder form (70-100 mesh)
- Polythene bags, HDPE bags, cardboard cartons
- Growth materials - include Manital, Sucrose and chemical nutrients.

Quality Control: As the products being living microorganisms, the quality checks up, batch-wise certification even if it is internal is highly essential as per BIS standards. Each unit should have following facilities:

- Adequate microbiological lab and qualified microbiologist
- Sampling and testing at various stages of production, including the quality of raw materials
- Quality certification
- Storage of the products in cooler places till they are sold to farmers
- Aseptic conditions, cleanliness and contamination free production lines and housing
- Automatic and closed systems

As per BIS specifications, certain tests are required to be conducted, like number of cells, colony character, reaction etc. Cell number at the time of manufacture should not be less than 10^8 and 10^7 per gram of carrier material, respectively for *Rhizobium* and *Azotobacter.* Similarly, the number of cell count and permissible contamination at expiry dates are also specified.

Process involved in biofertilizer production: The manufacturing process involves:

1. Selection of suitable strain of the organism
2. Mass multiplication
3. Mixing of the culture with carrier material and packing

The steps involved are as follows:

Culture selection and maintenance: The pure mother cultures of

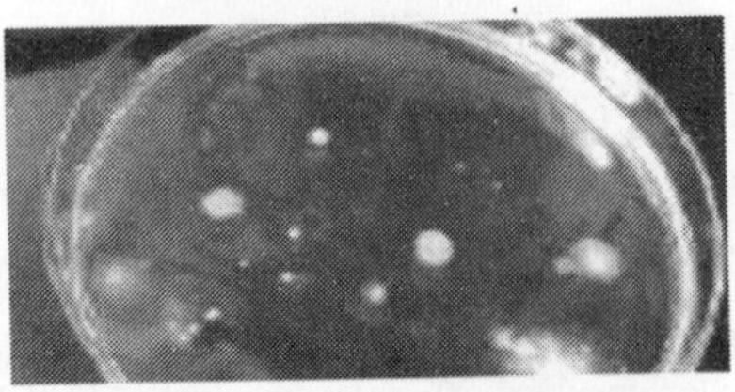

Colonies of various types of soil bacteria growing on a culture media in a Petri dish

various strains are being maintained in Agricultural Universities, ICAR institutions, Regional Centres of Organic Farming labs, etc. There are international sources of supply also like NifTAL, IRRI etc. The mother culture in test tubes of desired strain can be purchased from the identified sources. They have to be further sub-cultured and maintained purely for mass production by adopting standard techniques under the supervision of trained microbiologist.

Culture augmentation: In the next stage the culture has to be mass multiplied in two levels namely (i) at primary level using shakers in flasks and (ii) at secondary stage multiplication in fermenters.

The important factor in this is the preparation of growing medium in which the culture is mass multiplied. There are standard media on which information is available in case of Rhizobium from published sources like Norris & date, ISI approved etc. Similarly composition for growth media is available for other cultures.

After the media is formulated and sterilized in fermenter, it is inoculated using the shorter cultures multiplied in the flasks at definite ratios usually 5%. The bacteria growing medium are called broth and it is continuously aerated by passing sterile air from compressors. After about 3-4 days fermentation period, the broth will be ready for packing in a carrier material. At various stages the quality is tested by drawing samples.

Carrier sterilization: While the broth is getting ready in the fermenter the carrier material, which is usually the carbon source for the cultures to survive, is sterilized in autoclaves and kept ready for mixing the broth. Peat imported from countries like U.S., Australia is reported to be the best source of carrier material. However, as it is costly, lignite is used extensively in India and other developing countries. The carrier is either sterilized in bulk or it is packed and then the packets are sterilized.

Mixing and packing: There are 2-3 alternatives depends upon the sophistication and automation of the unit.

- Under non-sterile system, the broth is harvested from the fermenter into sterilized carrier - the mixing is done manually under aseptic condition and packed in polythene bags of desired quantity.
- In a slightly upgraded method, the broth and sterilized carrier are mixed mechanically in a blender and the material is packed using semi-automatic packing and sealing machine. In a slightly modified method some units are packing by delivering desired quantities of carrier and broth simultaneously from separate pipe conveyance system in to the polythene bags.
- Under a completely sterile system the carrier is taken in autoclavable polypropylene bags and pre-sealed into which the broth from fermenter is directly injected with the help of dispenser. The injection hole is immediately sealed. The packets are kept in incubation room for about a week before transferring to store room.

Sterile system of packing using auto syringe and dispenser is recommended to be the best method and all new units should follow and adopt this system.

***Limitations and constraints*:** The major limiting factors include:

- Narrow genetic base of mother cultures and lack of efficient and virulent strains suitable to various agro-environments.
- Unsatisfactory carrier material with uniform and consistent good quality comparable to imported peat material.
- Contamination in broth mixing and packing stages, not using completely closed system of production.
- Unsatisfactory packing material which reduces shelf life.
- Unsatisfactory storing conditions, particularly during the

distribution period. Exposure to high temperatures and sunlight destroy the microbial culture. They should be preferably kept in cold storage conditions.

- Not employing properly trained microbiologist.
- Lack of quality controls and certification procedures.

3.1 Production techniques for *Rhizobium*

Isolation, authentication and strain collection, preparation of mother culture media and mass production are four important steps required in production of *Rhizobium.*

Isolation: *Rhizobium* is located in the root nodules of legumes (pulse crops). Roots of legume plant should be used for isolation of *Rhizobium* as per procedure given below:

- Roots should be thoroughly washed with distilled water and then healthy and pink nodules are selected without damaging them.
- Place undamaged nodule in 95% ethanol for 5-10 seconds and then wash it with sterile water.
- Make the nodule surface sterile by acidified mercuric chloride solution (0.1% w/v).
- Again wash nodules for 5-6 times in sterile distilled water.
- Now crush the sterilized nodule in petridish with the help of forceps in the presence of sterile water.

Authentication and strain collection: For confirming purity of strain following tests are essential:

- Grow the strain on standard growth media like YEMA (Yeast Extract Manitol liquid medium) and observe colony characters.
- Grow the strain on CRYEMA (growth media) and confirm its growth. *Rhizobium* is transparent, colourless and glistening in appearance. While contaminants such as *Azotobacter* is pink and *Pseuedosomanas* and *Achromobacter* are red in appearance.

- Observe the fast or slow growth in BTB (Bromothymo Blue).
- Do gram reaction test. It should be -ve.
- Do GPA (Glucose pentone Agar) Test: Not favour growth of most rhizobia but many contaminants readily grow.
- Nodulation test should confirm effectiveness of *Rhizobium.*

Pure and effective strain of *Rhizobium* is selected for biofertilzer production.

Preparation of starter mother culture media (Broth): *Rhizobium* is heterotropihic bacteria hence utilise mono and disaccharides readily. Mannitol is useful as it is source of carbon and energy. The important culture media are AMD medium, Burton medium, ISI medium, MSM medium and YEM, etc. An example of typical media is: Glycerol (10 ml), Yeast extract (0.4g), K_2HPO_4 (0.5 g), $MgSO_4.7H_2O$ (0.2g), NaCl (0.1g) and Broth (1 lit).

For mother culture (broth), the selected strain should be inoculated into yeast extract manitol liquid medium (YEM) and incubated for 2-7 days at 28°C+2°C on a rotary shaker. The incubation period is dependent on the species.

Fermentation: The inoculated medium is sterilized in fermenter and inoculated with starter culture. The amount of inoculum culture to be transferred into the fermenter vessel is dependent on the size of the fermenter. Through sterile air and porous steel sperger, the broth is aerated continuously. The number of *Rhizobia,* 10^3 to 10^9 cell/ml, in broth is indicator to mix with carrier. Inoculum level is ranged from 0.1% to 5% but, in general, at the beginning of fermentation, culture media should provide 10^6 to 10^7 *Rhizobia*/ml.

Mean generation time for fast growing strains is 4 hours and 6-12 hours for slow growing strains and reach maximum viable number within 24-52 hours and 78-156 hours respectively.

3.1.1 Mass production method of Rhizobium

- Grind the carrier (peat or lignite/charcoal) and sieve it through 56-106 micron sieves. Sterilize the carrier in an autoclave and cool it. Determine the pH of carrier and neutralize (pH 7.0) it by adding $CaCO_3$.

- Prepare the starter culture (broth) as described above. Multiply it through orbital shaker or to the fermenter at 28±2°C temperature. Incubate the culture for 3-5 days and test it for its quality. The broth is aerated with a jet of sterile air at rate of 10-12 liter/hour if the material is in fermenter.
- Mix the culture broth with sterilized carrier. This is done when a cell count is about 10 cells/cc. Broth should be sufficiently added to occupy 40-50% of total water holding capacity of carrier.
- After mixing, the impregnated carrier is allowed to cure for 2-7 days at 28-30°C for further multiplication of *Rhizobium* in carrier.
- Impregnated carrier is then packed in the polythene bags. It should be stored at 15°C in cool and dry place before dispensing.

A schematic flowchart of mass production of *Rhizobium* is given below:

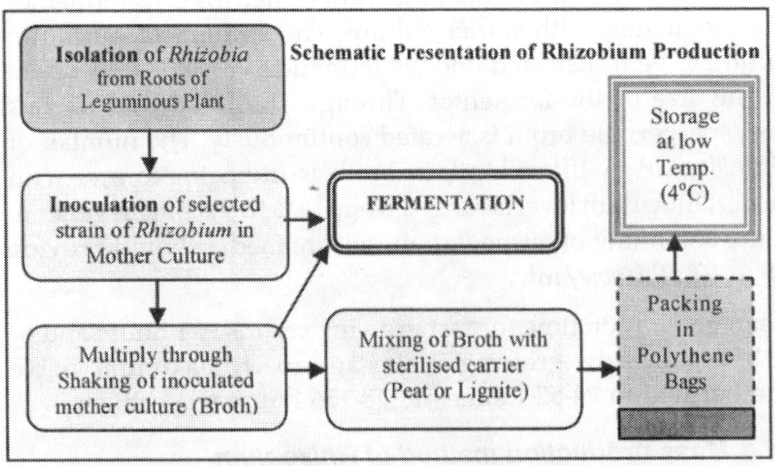

3.2 Production techniques for *Azotobacter*

Preparation of starter culture: Jensen's medium, Burk's medium, Waksman No-77 broth, Ashby's medium are important media for production of *Azotobacter.*

Ingredient required for preparation of *Jensen's medium* and *Ashby's medium* is given here:

Jensen's medium: Sucrose (20 g), K_2HPO_4 (1g), $MgSO_4.7H_2O$ (0.5g), NaCl (0.5g), FeSO4 (0.1g), $CaCO_3$ (2g), Agar (15g) and distilled water (1 lit).

Ashby's medium: Manitol (20 g), K_2HPO_4 (0.2g), $MgSO_4.7H_2O$ (0.2g), NaCl (0.2g), K_2SO_4 (0.1g), $CaCO_3$ (5g), Agar (15g) and distilled water (1 lit).

Azotobacter chrococcum is mostly used for production of *Azotobacter* biofertilizer. There are two methods of isolation of *Azotobacter* namely, soil dilution method and direct method.

Soil dilution method:

- Take 10 gm of soil samples and mixed it thoroughly with 100 ml of sterile distilled water.
- Prepare a suitable nitrogen free agar media (e.g. Jensen's medium) and pour into sterile Petridishes and cool it.
- Incubate the plates at 30°C for about 3-4 days. After incubation, soft, milkey, mucoid colonies of *Azotobacter* will appear on the plates. The older cells only show black to brown pigmentation.

Direct Method: In this method, soil is spread directly on the pertidish containing nitrogenous free Agar medium and incubated at 28°C. After for 3-4 days of incubation, colonies of Azotobacter are developed.

3.2.1 Mass Production

1. Pulverize the carrier (peat or lignite/charcoal) in a grinder and sieve it through 106 micron sieve.
2. Determine the pH of carrier and neutralize it by adding $CaCO_3$. Neutralization is not required in case of charcoal.

3. Sterilize the carrier in an autoclave and cool it.
4. Prepare the starter culture and multiply through shaker culture or in fermenter.
5. Mix the culture broth with carrier and maintain about 40% moisture content.
6. Finally pack it in polythene bags of medium density or low density (0.089-0.038 mm gauge) and seal it after removing the air.

3.3 Production techniques for *Azospirillum*

Isolation, purification, preparation of mother culture media and mass production are four important steps required in production of *Azospirillum*.

Isolation: *Azospirillum* belongs to the family spirillaceae. *Spirillum lipoferum(=Azospirillum)* is a common root used for production of *Azospirillum* biofertilizer. The isolation technique of *Azospirillum* is described below:

- Prepare a nitrogen free bromothymol blue (NFB) agar medium with following composition (g/lit):

  Bromothylmol blue-0.5% alcoholic solution (2ml), Malic acid (2g), K_2HPO_4 (0.5g), KOH (4g), $FeSO_4.7\ H_2O$ (0.05g), $MnSO_4.7H_2O$ (0.01g), $MgSO_4.7H_2O$ (0.01g), NaCl (0.02g), $CaCl_2$ (0.01g), Na_2MoO_4 (0.002g), Agar (1.75 g) and distilled water (1 lit). The pH (6.6-7.0) should be adjusted using NaOH.
- For isolation of *Azospirillum,* small pieces of washed roots preferably from rice plant (0.5 cm pieces) or small samples of soil (few micrograms) preferably from rice field are placed on an agar medium.
- In case of roots, cut the roots into small size of about 1 cm using stainless steel blade. Sterilize the small roots with 1% chloranium-T solution for about 2-5 minutes or 0.1% mercuric chloride for 1 minute. Wash the roots several times with sterilized water and phosphate buffer (pH 7.0).
- Incubate the tube for about 3-5 days at 28-30°C.

- After incubation period, observe surface pellicles and change of the colour of the medium from light yellow to blue which confirm the growth of *Azospirillum* on the medium.

Purification: Prepare malic acid medium with 1.5% Agar. Pour the medium into sterile petridishs. Collect the culture from tube and place over solidified medium. Incubate plates and observe the colour change. If the medium colour changes from light yellow to blue and the organism show its active spiral movement under microscope, it confirms the purity of *Azospirillum* in the medium.

3.3.1 Mass production technique of Azospirillum

Preparation of starter culture: Okon's medium, Semi-solid malate medium, etc. are important media for production of *Azospirillum*.

Ingredient required for preparation of *Okon's medium, NFB medium* and *Semi-solid malate medium* is given here:

Okon's medium: Sucrose (20g), K_2HPO_4 (6g), KH_2PO_4 (4g), $MgSO_4.7H_2O$ (0.2g), NaCl (0.1g), $FeSO_4$ (0.1g), $CaCl_3$ (0.02g), $FeCl_3$ (0.01g), $Na_2MoO_4.2H_2O$ (0.002g) and distilled water (1 lit). The pH (6.8) should be adjusted using NaOH.

Semi-solid malate medium: Malic acid (5g), K_2HPO_4 (0.5g), $MgSO_4.7H_2O$ (0.2g), NaCl (0.1g), $CaCl_2$ (0.02g), $MnSO_4.7H_2O$ (0.1g), KOH (4.5g), Biotin (0.1g), Fe-EDTA (4 ml), $Na_2MoO_4.2H_2O$ (0.002g), Bromothylmol blue (3ml) and distilled water (1 lit). The pH (6.8) should be adjusted using NaOH. Then add 1.5% agar.

After purification, *Azospirillum* strains are kept on Agar slants containing *Okon's medium* for 3-10 days. Culture is then transferred to large flask containing sterile broth and allows to grow for 3-5 days. Now large quantities of broth is prepared in production fermenter and sterilized. The pH should be maintained between 6.5 to 7.0 and broth is cool down at 30ºC.

Starter culture (from flask) at the rate of 1% of volume of broth is mixed in the production fermenter. Aeration is provided as per the size of the fermenter to get a maximum production of *Azospirillum*. At the end of fermentation, the broth culture should be checked for purity. Fermented broth culture should be mixed with sterilized carrier (peat/lignite) and maintain moisture to 40-45%. The product should be packed in polythene bags and sealed it after removing the air from polythene bag.

3.4 Production techniques for *Phosphobacterin*

The bacterial species belonging to the genera *Bacillus* and *Pseudomonas* and fungi species belonging to the genera *Penicillium* and *Aspergillus* possess the ability to bring insoluble phosphates in soil into soluble forms by secreting organic acids. Bacterial cell size ranges from 1 to 2 mm. Rod shaped *Bacillus* is gram +ve and *Pseudosomonas* is gram –ve bacteria.

For *Phosphobacterin* production following procedure should be adopted:

- Grow selected bacteria belonging to the genera *Bacillus* and *Pseudonomas* in Pikovskaiyas broth for 7-18 days at 28°C (±2°C).
- Mixed the broth in soluble sterilized carrier (peat soil).
- The mixture is cured for a week at the same temperature (28°C (±2°C) in a large tray and cover it properly with similar empty tray.
- After one week the inoculants will be ready for packing in a plastic bag (250-300g). Packet should be sealed and stored at 15-20°C until use.

The bacteria like *Bacillus polymyxa, Bacillus circulans, Aspergillus awamori, Pseudonomas striata,* etc. are important for increasing the yield of rice, wheat, maize, etc. Phosphate solublizing bacteria (PSB) can be recommended for all crops. PSB can solublise 20-30% of insoluble phosphates and increase the crop yield by 10-20%. PSB can be used like other biofertilizers.

4. Economics of the Biofertilizer Production

Based on the various techno-economic parameters, the economics of the project have been worked out. The expenditure includes the cost of raw material, transportation and power, fuel packing distribution, wages and salary, repairs and maintenance, insurance, advertisement, overheads, etc. and income includes sale of biofertilizers. The cost of equipments and development of infrastructure is not considered as they are permanent assets of the unit. For the model 150 TPA the relevant techno-economic parameters are furnished below:

I. Outlay of 150 TPA Bio-fertilizer Unit

Sl. No.	Particulars	Quantity
1	Land with Fencing and Gate in the front	2000 sq.mt.
2	Civil Structures	
	Sterilisation	525 sq.ft.
	Incubation and Quality Control	350 sq.ft.
	Maturation	500 sq.ft.
	Carrier Preparation and Enrichment	500 sq.ft.
	Storage	400 sq.ft.
	Staff / Stores	725 sq.ft.
3	Plant and Machinery	
A	Lab for Culture Maintenance & Quality Control	
	Laminar Air Flow	1 No.
	Laboratory Autoclave	1 No.
	BOD Incubator	1 No.
	Hot Air Oven	1 No.
	Microscope with CTV	1 No.
	Water Distillation Assembly	1 No.
	pH Metre	1 No.
	Balances	2 Nos.
	Colony Counter	1 No.
	Glassware	As per requirement
B	Preparation, Mixing & Packing	
	Orbital Shakers	2 Nos.
	Fermenter - 300 lit	3 Nos.
	Autoclaves (Horizontal)	1 No.
	Autoclaves (Vertical - Big)	1 No.
	Autoclave (Vertical - Small)	1 No.
	Steam Generator - 200 lit & Compressor	1 No.
	Demineralising Plant	1 No.

Sl. No.	Particulars	Quantity
	Hot Air Oven	1 No.
	Steam circulation system (Pipes & Insulation) semi- automatic mechanisms for broth injection and packing carrier material or fully closed system with two dispensers	1 Set
4	Other Fixtures	
	Fridges	2 Nos.
	Air conditioner - 2 Tons	2 Nos.
	Air conditioner - 1 Ton	2 Nos.
	Furniture and Fixtures	As per requirement
	Electrical Installations for 70 HP	As per requirement
	Well and Pump sets, Overhead Tank	As per requirement
	Generator for 70 HP	1 No.
5	Vehicles (LMV) & Jeep	1 No each

II. Income and Expenditure of 150 TPA Biofertiliser Unit

(Rs. in lakh)

Sr. No	Particulars	Yr 1	Yr 2	Yr 3	Yr 4 onwards
1	Installed Capacity	150 MT	150 MT	150 MT	150 MT
2	Capacity Utilisation	30%	40%	70%	90%
A	INCOME				
	1 kg. Packet	10.238	13.650	23.888	30.713
	5 kg. Packet	3.983	5.310	9.293	11.948
	Loss due to Contamination	0.711	0.948	1.659	2.133
B	Total Income	13.509	18.012	31.521	40.527
C	EXPENDITURE				
	Power	0.278	0.417	0.833	1.250
	Medium Preparation	0.223	0.334	0.667	1.000
	Wages	0.223	0.334	0.667	1.000
	Water Charges	0.043	0.067	0.133	0.200
	Packing Material	1.575	2.100	3.675	4.725
	Carrier Material	0.556	0.834	1.667	2.500
	Rent, Rates, Taxes	0.135	0.180	0.315	0.405
	Administrative Expenses	1.351	1.801	3.152	4.053
	Marketing Expenses	2.026	2.702	4.728	6.079
	Clearing-Forwarding	0.068	0.090	0.158	0.203
	Mother Culture	0.240	0.240	0.240	0.240
	Interest (Bank loan)	0.708	0.711	0.716	0.725
D	Total Expenditure	7.426	9.810	16.951	22.380
H	Profit	6.083	8.202	14.57	18.147

5. Technical Aspects of Biofertilizer

5.1 Methods of placement of biofertilizers

Biofertilizers are usually inoculated by seed pelleting or seedling dipping or through soil applications depending on the nature of the crop. The *Rhizobium* biofertilizers are usually inoculated as seed pelleting. When the seed germinates, the root coming from the seed coat gets infected and later, the nodules are formed in the legume which fixes atmospheric nitrogen in the symbiotic process. On the other hand, biofertilizers like *Azotobacter* & BGA grow and fix atmospheric nitrogen free livingly. However, *Azospirillum* likes to grow very close to the root rhizosphere which are also called as associate symbionts which grow and colonize the root zones of the beneficiary crop. Many of the biofertilizers which are used for perennial are added to the crop at a later stage which is mixed along with well composted farmyard manure (FYM). Usually *Azotobacter* and *Azospirillum* are used through seedling dipping for transplanted crops, while phosphorus solublising bacteria (PSB) are used as soil application. Although, the package of practice differs from crop to crop, however, the biofertilizer could be used in any of the methods.

Seed treatment for broadcasting: One packet of 200 g biofertilizer is mixed in 500 ml of water to prepare slurry and the slurry is mixed with about 10-12 kg of seeds and is air dried under the shade. Soon after the drying, the biofertilizer coated seeds are broadcasted.

Seedling dipping for transplantation: In this method 1 kg of biofertilizer, which is sufficient for seedling needed for one acre, is mixed with 5 litres of water in a bucket and mixed thoroughly. The roots of seedling bundles are dipped in the biofertilizer suspension for about 30 minutes and taken out. The seedlings root coated with biofertilizers are transplanted immediately.

Soil application: In soil application, 4 to 5 kg of desired biofertilizer is mixed with 100 kg of well decomposed FYM and mixed thoroughly, so as to get a uniform mixture. Biofertilizer mixed FYM is applied to the soil after 15 to 20 days of seedling transplantation of vegetable crops or one month after broad casting of seeds or FYM is applied in the pits before transplantation or in dragged ring, so as to ensure the placement near the root zone of perennial crop. Soon after biofertilizer is applied, it should be covered with soil, so as to avoid exposure to direct sunlight. Placement of biofertilizers is another approach for its successful colonization. Since, the microbes like to grow and proliferate at the rhizosphere zone of the crop root, the added inoculants should be placed close to the root zone (Rhizosphere) so that, the mutual requirements of microbes and the beneficiary crop are met and the inoculated microbe could easily be established in the root rhizosphere.

The biofertilizers should be stored in cool place and should not be exposed to direct sun light. After, the biofertilizer is applied it is covered immediately by soil under moderate moisture. The efficiency when applied to soils is limited by several factors; most important of them are- drought and high summer temperature, water logging, unfavourable soil pH, antagonism from other organisms and nutrient deficiency.

5.2 Mode of action

The mode of action depends on the species of the organism. Some agents like *Rhizobium* cultures enhance N fixation in legumes by imparting effective modulation as they are symbiotic bacteria living in association with leguminous plants.

There are free living bacteria like *Azotobacter* when applied to soil, enhance the N availability. There are certain other organisms which act on the soil minerals and dissolve the native nutrients like phosphorus which is otherwise not readily soluble.

5.3 Critical factors for effectiveness of biofertilizers

The critical factors which are responsible for the effectiveness of a particular biofertilizer are as follows:

- Suitability of the species to the target crop
- There are specific strains of *Rhizobium* for different leguminous species like Cowpea, Redgram, Soybean, Alfalfa etc. Biofertilizer of specific culture should be used for specific crop.
- Identification of strains as suited to the agro-eco system, particularly the soil pH and moisture conditions. Specific strains as suited to a particular soil and environmental conditions are usually identified and pure mother cultures are maintained in research laboratory for supply to the commercial manufacturers, e.g. germplasm of *Rhizobium* cultures is maintained at IARI, New Delhi.
- The aseptic conditions of manufacturing, the cell count of living organism present in the carrier material, purity and level of contamination.
- The conditions of carrier material in which the culture is packed and the quality of the packing material, which determine the shelf life.
- The conditions in which the packed materials are stored, distributed and kept with the farmers before it is applied.
- Soil conditions particularly pH, organic matter content and moisture level and agronomic practices.

5.4 Level of benefits

The benefits usually obtained will not be as visible as that of mineral fertilizers. Biofertilizers can add nitrogen from 20 kg/ha to 70 kg/ha depending upon the optimum conditions. Pastures and forages respond more than grain crops. The yield increases usually range around 10-35%. However, in the vast areas of low input agriculture and in the context of imparting sustainability to crop production at reduced chemical pollution like organic farming, this product will be of much use.

Application of biofertilizersshould not be viewed from the only angle of nutrient supply to the crops. They add life to the soil rendered sterile by the excess use of chemicals, etc. Some of them possess growth promoting substances and also reduce the incidence of certain diseases. These inputs are crucial if some one would like to take up organic farming.

Chapter **5**

Azolla, BGA & Mycorrhiza (VAM) Production

Azolla is a water fern commonly found floating in idle pond, tanks, shallow ditches and channels. Blue green algae (BGA) living in the epidermal cavity of the lower side of the leaf of Azolla in symbiotic association. The algae fix atmospheric nitrogen in presence of sunlight for Azolla and in exchange the plant provides home and food to the algae. This symbiotic association of *Azolla pinnata* and *Anabeana azolla* is termed as *Azolla-Anabeana Complex*. As this complex fixes atmospheric nitrogen, it has great potentiality for use in organic agricultural fields as an alternative to chemical nitrogenous fertilizers.

Mycorrhizae means "fungus root" associated with the fine roots of most plants. There are hundreds species of fungi which function as mycorrhiza; most are basidiomycetes, the class of fungi which form mushrooms. Mychorrhiza is the symbiotic association between fungal mycelia and roots or rhizomes of higher plants from pteridophyta, gymnosperms and angiosperms. Mycorrhiza increases the surface area of the root system for better absorption of plant nutrients from soil and keep plants healthy and fast growing.

Azolla, BGA, Mycorrhiza and beneficiary crops

Name	Benefits	Beneficiary Crops
Blue Green Algae	Fix 20 -30 kg N/ha	Rice (Lowland or flooded)
Azolla	Azolla can give biomass up to 40-50 tonnes and fix 30-100 kg N/ha	Rice (Lowland or flooded)
Mycorrhiza (VAM)	30-50% yields increase Enhances uptake of P, Zn, S and water by plant.	Fruit trees, crops and ornamental plants

1. Production Technique of Azolla

Currently, there are seven *Azolla* species identified world wide - *Azolla caroliniabna A. filiculodies, A. mexicana, A. microphylla, A. ruba, A. pinnata and A. nilotica.* Azolla has a very high rate of vegetative propagation with doubling time of two days in favourable conditions. Because of its rapid multiplication, it has unique property to retain significant amount of atmospheric nitrogen. Fresh Azolla can be used in flooded rice ecosystem and composted Azolla in any crops.

1.1 Ideal Conditions for growth and multiplication of Azolla

- Azolla grow well within 14-32°C temperatures with water depth of 8-10 cm.
- The suitable pH for growing Azolla lies between 6.0 to 8.0
- The occurrence of standing water and application of phosphorus (organic source) are the two important requirements for rapid growth of Azolla.

1.2 Method of growing Azolla in nursery

- Azolla nursery should be prepared simultaneously with the paddy nursery. The area of Azolla nursery should be half of that of paddy nursery.
- Select the proper nursery field and divide in plots (20 m x 2 m) with bunds and irrigation channels. The nursery should be well prepared and leveled.
- Apply 250 g rock phosphate/plot (acidic soil) and irrigate the field for dissolution of phosphates.
- After 10 days of Rock Phosphate application, each plot should be flooded with irrigation water up to a depth of 8-10 cm.
- Prepare a cattle dung by mixing 10 kg fresh dung in 10 litres of water and spread it on each plot.
- Apply 8-10 kg fresh Azolla/plot.
- When temperature is favourable, depth of water should

be maintained. Azolla grow to double by weight in 3-4 days.

- Azolla can be harvested after 15 days of application and introduced in the rice field as source of nitrogen.
- About 40-50 kg fresh Azolla can be harvested from each plot. The same Azolla can be used for further production in the nursery field.

Composting: For composting, Azolla can be collected every week from nursery and kept in layers on the side of nursery bed. Within 10 days, the Azolla will decompose and can be used as compost in any crop.

When Azolla covers the whole field, it indicates that it has grown to the extent of 10 tonnes per hectare and is equivalent to one ton of dried or decomposed Azolla. It adds about 30-40 kgs of nitrogen (equivalent to 65-85 kgs of urea) in the soil. It also adds approximately 5-10 kgs of phosphorus and 25-50 kgs of potassium besides micronutrients.

Azzolla production in Nursery bed

Azolla contains high protein content and increases the organic matter of the field. Cost of production of Azolla is very less and it can be grown without incurring any cost. It also control the weed infestation in rice crop as it is grown as dual crop or inter crop or green manure crop.It also serves as a good protein rich fish and poultry feed.

2. Production Technique of Blue Green Algae (BGA)

Blue green algae are also called as 'Cynobacteria'. Predominant BGA are species of *Aulosira, Anabaena, Nostoc, Cylindrospermum, Gloeotrichia* and *Aphanothece.* Most BGA are dominant in the moist soil and *Aphanothece* is found in all water regimes whereas

Aulosira preferred clean standing water.

Readymade spore or culture of BGA is available at ICAR Centre of almost all the states of India. It is important to select fast growing and competitive strains as seedling material. However for production of culture, BGA can be extracted from soil by isolation and purification as discussed below:

Isolation: BGA can be isolated from rice fields. The procedure for isolation BGA is given below:

- Prepare a culture media (Fogg's medium) with following composition:

  KH_2PO_4 (0.2g), $MgSO_4$ (0.2g), $CaCl_2$ (0.1g), Na_2MoO_4 (0.1mg), $MgCl_2$ (0.1mg), H_3BO_3 (0.1mg), $CuSO_4$ (0.1mg), $ZnSO_4$ (0.1mg), Fe-EDTA (1ml) and distilled water (1Lit).

- Take 100 ml sterilized culture media in a 250 ml volume conical flask and add 5 g soil samples collected from rice field.
- Shake the mixture for few minutes and then incubate in growth room without any intermittent shaking. The growth room should be provided with 500 lux illuminated temperature control system.
- After the formation of algal growth, it should be examined under microscope for contamination. Take purified algal clump and introduce into the test tube containing 5 ml sterilized water. Homogenize the content by vortexing.
- Now dilute the content to 10^{-1}, 10^{-2} and 10^{-3} by sequence. Take 1 ml of diluted suspension in a plate from 10^{-1}, 10^{-2} and 10^{-3} dilution by pourplate technique.
- Incubate the petriplates upside down in a growth room once the agar medium has solidified.
- After appearance of colonies, inoculate it in a agar slants. Slants may further be allowed to grow and examined for purity of algae.
- Repeat the dilution and plating procedure once again.

- If unialgal found, purification of mixed culture of BGA should be done by repeating plating.

Process of Purification: Take nutrient broth in test tubes containing beef extract (3g), tryptone (5g), and distilled water (1 lit) with pH 7.0. Inoculate the purified algal suspension of test tubes in bacterial incubator at 28°C for 2 days. No turbidity confirms the absence of bacterial contamination.

2.1 Mass Production technique

1. Prepare shallow trays (2mx3mx23m) of galvanized iron sheet or permanent tanks.
2. Place a layer of 10 kg soil in tray and mix it well with 200 g each of Rock phosphate and of vermicompost. To prevent insects, add neem oil or any other suitable biopesticides.
3. Add water up to 5-15 cm to the tray and test the pH. If acidic, correct it by adding lime.
4. After the soil settles down, sprinkle the algal culture and sawdust and keep the unit in open air completely exposed to sun.
5. After about 8-10 days, a thick algal mat will be formed. If daily rate of evaporation is high, add water intermittently and when the algal growth becomes sufficiently thick, stop watering.
6. Allow the water to evaporate completely in the sun.
7. The dry algal cracks into flakes, which is collected and stored in gunny bags for field use.
8. Fill the tray with water and add small amount of the dried algal flakes for further inoculum.Continue the above process till the soil in the tray is exhausted. About 3-4 harvests can be taken from one soil preparation.

A single harvest of surface algae from one tray usually will give 1.5-2.0 kg of material, which is enough to inoculate 0.5 ha of rice field. Trials conducted so far show that a continuous application of these algae atleast for three consecutive seasons results in an appreciable population build-up of the inoculated strains which is reflected in crop yield.

3. Field I noculation and Multiplication of VA Mycorrhizae

Mycorrhizas are highly evolved, mutualistic associations between soil fungi and plant roots. The partners in this association are members of the fungus kingdom (Basidiomycetes, Ascomycetes and Zygomycetes) and most vascular plants. Mycorrhizae can benefit plants by enhancing the nutrient absorbing ability of roots. An individual plant may have several different mycorrhizae associated with its roots.

Mycorrhizae are especially important in dissolving nutrients such as phosphorus from mineral particles and carry them to the plant. This enhancement of nutrient uptake is a result of the extensive system of hyphae and mycelia (thread-like filaments of the mycorrhizal fungus) that pervade soils. They function like root hairs but are much more far reaching. They also impart resistance to the plants against the soil borne fungal pathogens and nematodes and also make micronutrient available to the plants.

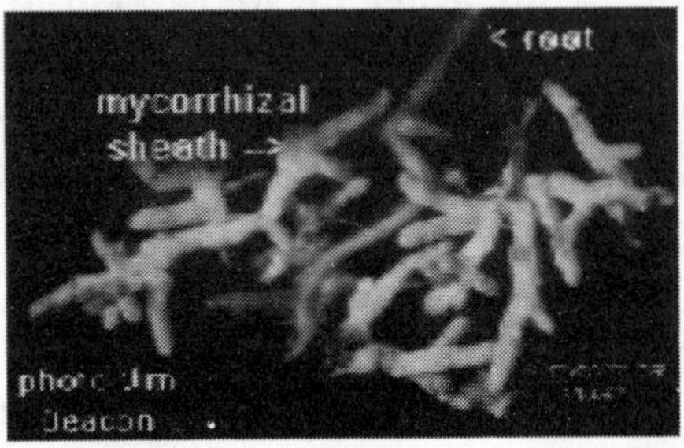

3.1 Type of mycorrhizae

Considering the morphological and anatomical features the mycorrhizae are grouped into of two types: (i) Endomycorrhizae and (ii) Ectomycorrhizae.

Endomycorrhizae: In this case, the fungal hypae are present inside the root. Some endomycorrhizae are also called as Vesicular Arbuscular Mycorrhizae (VAM) fungi. Vesicular Arbuscular Mycorrhiza (VAM) is the most abundant kind of mycorrhiza described as 'a universal plant symbiosis'. VAM are associations where Zygomycete fungi in the Glomales produce arbuscules, hyphae and vesicles within roots. Spores are formed in soil or roots. These associations are defined by the presence of arbuscules. Fungi within roots spread by linear hyphae or coiled hyphae.

Endomycorrhizae can not be cultivated in their pure forms in laboratory since they are obligate parasites. For their culture and maintenance, a living host like maize, sorghum, millets, cowpea, clover, etc. is required. VAM are used for the annual plants such as cereals, pulses, oilseeds, vegetables, etc. Almost all the plant species are infected with the VAM except cruciferous plants.

The important species of five genera are- *Glomus mosseae, Glomus fasciculatum, Gigaspora nigra, Acaulospora scrobiculata, Sclerocystis clavispora and Endogome increseta.*

Ectomycorrhizae: Fungal mycelium forms a covering on the surface of roots. Root growth is stunted in infected plants. Ectomycorrhizae can be cultivated in the laboratory in artificial media and the culture can be used for the inoculation just like rhizobium and other microorganism. Ectomycorrhizae are basically used for inoculation of the tree plants.

3.2 Inoculation methods of mycorrhizae production

3.2.1 Vescular Arbuscular Mycorrhizae (VAM)

1. Maize, sorghum, onion and garlic crops favoured hosts for maximum number of mycorrhizae species in their roots. The living roots of these crops should be separately collected after harvesting of above crops under moist condition.
2. They should be washed and cut into fine pieces (1-2 cm) with the help of chaff cutter. These root pieces are to be applied in the field

as soil inoculants of VAM fungi.

3. About 25-30 kg of root pieces can be used for inoculating the one hectare of land during crop season. While adding the root pieces in the field, phosphatic fertilizers (Rock phosphate) @ 10kg/ha may be used to initiating the activity of roots.

Soil samples collected from rhizosphere of maize, sorghum, onion and garlic crop can also be used as a source of VAM fungi. The soils collected from such field can be applied in the field at the time of sowing of new crops either in furrow or following broadcasting method.

VAM fungi can also be isolated from soil in laboratory by following methods:

1. Collect 10 g of soil samples from the rhizosphere of maize, sorghum, onion or garlic plants and mix in 40 ml of water in graduated cylinder (50 ml).
2. Shake the soil mixture for two minutes on shaker to allow the spores to float and accumulate at the top.
3. The aqueous suspension is slowly decanted to transfer the spores into a funnel mounted with a 988 micron meter wire mesh and filter paper below it.
4. Large particles are retained on the wire mesh and the spores are collected on the filter paper.
5. The process is repeated twice with the soil residues in the graduated cylinder. Large particle should be eliminated in every process.
6. Filter paper is examined under the stereoscopic binocular microscope. The spores of VAM should be picked up by the needle.

VAM Spores on filter paper

7. Precaution should be taken that plant roots or soils collected should not be from the fields treated with chemical pesticides or fertilizers.

There are two principal ways of ensuring that the benefits in terms of crop production are obtained from mycorrhizal associations by:

- inoculating with selected efficient mycorrhizal fungi, and
- promoting the activity of effective indigenous mycorrhizal fungi by proper cultural practices.

Cultural practices that increase the activity of indigenous VAM fungi are: reduced tillage, crop rotations, cover crops, and phosphorus management. This promotes rapid colonization of a new crop and enhances early season mycorrhiza-mediated P uptake.

Chapter 6

Plant Disease and Pest Management

Insect-pests and diseases strike primarily at weak and improperly nourished plants. Pest problem is one of the major constraints for achieving higher production in agriculture crops. Indialoses about 30% of its cropsdue to pests and diseases each year. The damage due to these is estimated to be Rs.60,000 crores annually.

The problem of insect-pest and disease is acute in case of all the crops and especially so in case of commercial crops. The use of insecticides, pesticides and fungicides has increased manifolds during the past 3-4 decades with the introduction of intensive cropping. The average consumption of pesticides in India is about 570 gms per hectare as compared to developed countries like Japan, Thailand and Germany where the consumption rate is 11 kg, 17 kg and 3 kg per ha, respectively. Though the average quantum of pesticides usage in India is low, the damage caused due to their indiscriminate usage and poor quality maintenance is alarming.

Pesticides, fungicides or chemicals are meant to control harmful pests such as insects, nematodes, diseases, weeds etc. However, excessive use of pesticides and fungicides not only leave residues in soil, water and air but also have adverse effects on the non-target organisms such as pollinators, parasitoids, predators and wild animals. This has adversely affected the ecological balance resulting in pest resurgence, development of resistance in the pest species, diseases and environmental pollution.

In view of the several disadvantages associated with the unscientific use of pesticides in agriculture, there is an urgent need for minimising the use of chemical pesticides in the management of insect-pests. Growing public concern over potential health hazards of synthetic pesticides and also steep increase in cost of cultivation/low profit making by farmers has led to the exploration of eco-friendly organic farming.

One of the objectives of organic methods is to grow crops which naturally resists the onslaught of pests and diseases. Pest and disease management in organic agriculture consists of a range of activities that support each other. Most management practices are long-term activities that aim at preventing pests and diseases from affecting a crop. Management focuses on keeping existing pest populations and diseases low. Control, on the other hand, is a short-term activity and focuses on killing pest and disease. The general approach in organic agriculture is to deal with the causes of a problem rather than treating the symptoms, an aspect that also applies to pests and diseases. Therefore, management is of a much higher priority than control.

Management of soil tilth, moisture and nutrient status is the first step in effective pest and disease management. Crop rotations planted with intention of breaking life cycles of insects and pathogens, is a traditional means of pest control.

1. Preventive Measures for Disease and Pest Management

As many factors influence the development of pest and disease, knowledge about plant health and pest and disease ecology helps the farmer to choose effective preventive crop protection measures. This can be accomplished through the correct timing of management practices, a suitable combination of different methods, or the choice of a selective method.

Prevention and sanitation procedures in organic farming include post-season destruction of vines via tillage, burning or composting; removal of diseased plants and weeds; sterilization of plant stakes prior to re-use; and frequent cleaning of tools and implements to prevent transporting problems between fields.

Some important preventive crop protection measures are the following:

- *Selection of adapted and resistant varieties*: choose varieties that are well adapted to the local environmental conditions (temperature, nutrient supply, pests and disease pressure), as it allows them to grow healthy and makes them stronger against infections of pests and diseases.
- Selection of clean seed and planting material (certified seed and planting material from safe sources).
- *Use of suitable cropping systems*: diverse cropping systems, crop rotation, green manuring and cover crops.
- Use of balanced nutrient management: moderate N-fertilization and steady growth makes a plant less vulnerable to infection.
- *Input of organic matter*: improves soil fertility, stabilizes soil structure and supplies substances that strengthen the plant's own protection mechanisms.
- Compost can reduce disease problems due to the presence of microorganisms. They either competes with pathogens for nutrients, produce antibiotics that reduce pathogen survival and growth, or parasite on the pathogens. There is also an indirect effect on crop health.
- *Application of suitable soil cultivation*: regulates weeds that serve as hosts for pests and diseases.
- *Use of good water management*: flood irrigation and water on the foliage encourages pathogen infections.
- *Conservation and promotion of natural enemies*: provide an ideal habitat for natural enemies to grow and reproduce.
- *Selection of optimum planting time and spacing*: sufficient distance between the plants allows good aeration and reduces the spread of a disease.
- *Use of proper sanitation measures*: removes infected plant parts

from the ground to prevent the disease from spreading, eliminates residues of infected plants after harvesting.

2. Curative Crop Protection Methods

If all preventive crop protection practices fail to sufficiently prevent economic losses to the farmer, it may be necessary to take curative action. Curative action means controlling the pest or disease once it has already infested the crop. Several options exist in organic agriculture:

- Cultural and mechanical control with traps or hand picking.
- Biological control with natural predators or antagonistic microbes.
- Natural pesticides based on herbal preparations or other natural products.

Cultural Methods: Cultural control is the cheapest of all control measures. However, this technique is indirect and intangible. The cultural practices which can be gainfully adopted in different cropping systems to suppress certain pest species are tillage operation, field and plant sanitation, crop rotation, growing pest resistant varieties, trap cropping, adjusting the time of sowing, wider spacing, growing of barrier crops, water management. Other cultural practices also play a role. Important roles of cultural method are enumerated below:

- Orientation of rows to maximize air circulation helps reduce fungal disease problems.
- Suspending field activities when vegetation is wet with dew or rain limits the spread of disease, as does mulching to reduce direct soil contact and rain splash.
- Drip irrigation is preferred over sprinkler irrigation to reduce moisture and splash onto leaves and thus foliar disease occurrence.
- Solarization, or heating soils by tarping with clear plastic prior to planting, is a non-chemical soil treatment for suppression of diseases, nematodes, and other pests. As a

practical matter, however, its use is limited to small-scale operations.

Mechanical methods: Hand picking of egg masses, gregarious larvae and sluggish adults, and their destruction helps in reducing pest populations in certain situations. Some of the effective mechanical methods of pest control are- pheromones and light traps, yellow sticky traps, nylon net, etc.

Botanical preparation: Many botanical insecticides have been known and used for hundreds of years. Botanical insecticides have different chemical structures and modes of action. Some of important botanical preparation being used in organic agriculture are- limonene and linalool, neem, pyrethrum, rotenone, ryania, sabadilla, horticultural oil, etc.

Microbial insecticides: The insecticidal crystal proteins (endotoxins) produced by the bacterium, *Bacillus thuringiensis kurstaki* are effective against lepidopteran pest species. These toxins are very specific in their action, easily biodegradable and being stomach poisons, safer to non-target organisms. Other microbial products that have been recently introduced for pest management are the insecticidal toxins produced by soil borne Actinomycetes organisms. The *A. vermectins* (abamectine) derived from *Streptomyces avermitillis* possess potent anthelminthic, acaricidal and insecticidal properties at very low doses. Like wise, Spinosyns (Spinosad), toxins obtained from another Actinomycetes organism, *Saccharopolyspora spinosa* exhibit very good insecticidal property against lepidopteran pests like diamondback moth, gram pod borer, etc.

Fungal pathogens like *Metarrhizium* sp., *Beauveria* sp., *Nomuraea rileyi*, *Paecilomyces farinosus* can be sprayed against caterpillars and other pests. *Verticillium lecanii* can be used against homopteran pests viz. scales and aphids. Another fungus, *Hirsutella thompsonii* has been found to be effective against citrus mite and coconut mite.

Antagonistic organisms: The various antagonistic organisms employed, as bio-control agents are *Trichoderma* spp., *Aspergillus* niger, *Gliocladium* virens, *Bacillus subtilis, Pseudomonas fluorescence, Paecilomyces lilacinus* and *Verticillium chlamydosporium,* Nuclear polyhedrosis virus (NPV) etc.

Insect Parasitoides and Predators: They are egg-parasitoids trichogramma, larval-parasitoids, Goniozus and Bracon, Pupal-parasitoids, Spalangia species, predacious green lacewings, Chrysoperla species, Predacious ladybird beetles, etc.

Mineral insecticides: Diatomaceous earth is a non-toxic insecticide mined from the fossilized silica shell remains of diatoms, (single-celled or colonial algae). It absorbs the waxy layer on insect bodies, abrades the skin, and dries out the insect.

Sulphur is the oldest known pesticide used as a dust, wettable powder, paste or liquid, primarily for disease control (e.g., powdery mildews, rusts, leaf blights, and fruit rots). However, mites, psyllids and thrips also are susceptible to sulphur. Sulphur is non-toxic to mammals, but may irritate skin or especially eyes.

3. Use of Bio-Fungicides and Bio-Pesticides

Fungicides: Fungicide options are limited in organic production; copper and sulfur-based products are the only labeled fungicides allowed in certain certification programs. Coppers are labeled for anthracnose, bacterial spot, early and late blight, gray leaf mold and septoria leaf spot. Sulfur is labeled for control of powdery mildew. Copper functions both as a fungicide and bactericide. Most formulations are allowable in organic certification. These include Bordeaux, basic sulfates, hydroxides, oxychlorides and oxides.

The use of copper fungicides in organic production is somewhat controversial. It is directly toxic at applied rates to some beneficial organisms, particularly earthworms and some soil microbes such as blue-green algae (an important nitrogen fixer) in many soils. Excessive use can also result in the buildup to phytotoxic (crop damaging) levels of copper in the soil. Thus, organic growers should monitor soil copper levels through regular soil testing.

Biological fungicides are a relatively new tool available to organic growers. Biological fungicides contain beneficial bacteria or fungi (microbial antagonists) which help to suppress pathogens that cause plant disease. For example, F-Stop™, registered as a seed treatment for tomatoes, contains a biocontrol agent called *Trichoderma viride sensu.* T-22G Biological Plant Protectant Granules™, registered as an in-furrow soil treatment on tomatoes and other vegetables, contains *Trichoderma harzianum, strain* KRL-AG2.

Weather data on canopy temperature, leaf wetness periods and other factors that affect the likelihood of disease occurrence are collected and are used to time fungicidal sprays for their optimum effect in growing seasons.

Biopesticides: Biopesticides are pesticides derived from such natural materials as animals, plants, bacteria and other microbes and certain minerals.

Three different categories of biopesticides are:

1. Microbial pesticides;
2. Plant pesticides;
3. Biochemical pesticides

Microbial pesticides consist of a microorganism (e.g. a bacterium, fungus, virus or protozoa) as the active ingredient. Microbial pesticides can control many different kinds of pests, although each separate active ingredient is relatively specificfor its target pest(s). The most widely used microbial pesticides are subspecies and strains of *Bacillus thurisngiensis,* or Bt. Each strain of this bacterium produces a different mix of proteins, and specifically kills one or a few related species of insect larvae.

Plant pesticides are pesticidal substances that plants produce from genetic material that has been added to the plant. For example, scientists can take the gene for the Bt pesticidal protein and introduce the gene into the plant's own genetic material. Then the plant, instead of the Bt bacterium, manufactures the substance that destroys the pest. Both the protein and its genetic material are regulated by EPA; the plant itself is not regulated.

Biochemical pesticides are naturally occurring substances that control pests by non-toxic mechanisms. Biochemical pesticides include substances, such as insect sex pheromones, that interfere with mating, as well as various scented plant extract insect pests to traps.

Biopesticides usually are inherently less harmful and affect only the target pest and closely related organisms, in contrast to broad-spectrum conventional pesticides that may affect organisms as different as birds, insects and mammals. Biopesticides often are effective in very small quantities and often decompose quickly, thereby resulting in lower exposures and largely avoiding the pollution problems caused by conventional pesticides.

4. Other Methods of Pest and Disease Control

Several natural remedies can also be employed by organic farmers. These include a wide range of products and practices like: compost watery extracts; hydrogen peroxide; sodium bicarbonate; foliar EM (effective microorganism); plant extracts (fermented nettele tea. Equisetum tea, comfrey tea); and biostimulants (seaweed, humates). The precise mode of action for many of these materials remains to be discovered. Of these, compost water extracts and hydrogen peroxide look promising for the control of tomato diseases like early and late blights. Compost extracts have proven effective for several vegetable diseases, including late blights. Compost extracts have proven effective for several vegetable diseases, including late blight of tomatoes.

Apart from plant extraction, there are some other natural pesticides that are allowed in organic farming. Although some of these products have limited selectivity and are not fully biodegradable, there are situations when their use is justified. However, in most cases, the desired effect is best reached in combination with preventive crop protection methods:

- Ashes: against soil-borne disease, ants, leaf miners, stem borers

- Slaked lime: against soil-borne diseases
- Clay: against fungal diseases
- Baking soda: against fungal diseases
- Soft soap solutions: against aphids and other sucking insects
- Light mineral oil: against various insect pests (harms natural enemies)
- Sulphuric acidic argillaceous earth: against fungal disease

The disease and pest management methods listed above can be integrated in to the organic farming systems depending upon the crops and the pest species. It is possible to keep the pest populations below economic injury levels by conserving existing natural enemies and adoption of suitable cultural, mechanical and biological methods. In addition, we have a large number of ITK (Indigenous Technical Knowledge) passed on from generation to generation, which can also be adopted along with other methods.

Further Readings :

1. Conversion to Organic Agriculture by A.K. Singh. International Book Distributing Co., Lucknow.

Chapter 7

Biopesticide Production

In organic farming, biological control of pests is one of the important means for checking pest and disease problems in almost all agro-ecological situations. Biopesticidesare living organisms which can intervene the life cycle of insect-pests in such a way that the crop damage is minimized. The agents employed as biopesticides, include parasites, predators and disease causing fungi, bacteria and viruses, which are the natural enemies of pests. Further, they complement and supplement other methods of pest control. On the other hand, these bio-agents can be conserved, preserved and multiplied under Laboratory condition for field release. Once these bioagents are introduced in the field to build their population considerably, they are capable of bringing down the 'targeted pest' population below economic threshold level (ETL).

Biopesticides are preferred in organic farming for the following reasons:

- No harmful residues
- Target specific and safe to beneficial organisms like pollinators, predetors, parasites etc.
- Growth of natural enemies of pests is not affected, thus reducing the pesticide application
- Environmental friendly
- Cost effective

The popularity of biopesticides has increased in recent years, as extensive and systematic research has greatly enhanced their effectiveness. Also, techniques for the mass production, storage, transport and application of biopesticides have been improved in recent years. Last decade has witnessed a tremendous breakthrough in this aspect, especially on standardization of production techniques of *Trichoderma, Gliocladium, Paecilomyces, Pseudomonas, Trichogramma,* NPV and Bacillus to use them against many insect pests and diseases.

1. Commercial Production of Bio-pesticides

Some of the local small-scale industries have already started production and marketing of *Trichoderma viride* (against few fungal diseases) and *Trichogramma* (against few insect-pest). They are able to meet the demand of only less than 1% of cropped area. There is a scope to enhance production and use of biological control agents in the days to come as the demand is increasing every year especially in the organic farms.

1.1 Basic requirements

Based on the field visits to biocontrol production units and in line with the technology and objective of biopesticides production, various facilities required for the successful implementation of such projects are indicated below:

Location of biopesticide units: In order to achieve optimum results, care needs to be taken to set up biopesticide facilities in areas which have appropriate climatic conditions. The production of biopesticides requires controlled climatic conditions especially temperature and humidity. Besides the climatic conditions, the proximity of the location to the market is also important. Also, the production should be located away from industrial and urban areas as air pollution can damage biopesticides.

Land: Land is required for construction of culture and rearing rooms, processing room, laboratory, office etc.

Building and civil works: Bio-pesticides production involves rearing of insects. Hence, the basic infrastructure to be created includes only the civil structures built in such a way as to provide environmental conditions suitable for rearing of insects. An estimated built up area of about 1000 sq ft is required for mass production of *Trichogramma, Chrysoperla* and *Cryptolaemous* beetles and about 2400 sq. ft. is required for production of NPV, *Trichoderma* and pheromone lures. Other utilities required are power, water and vehicle. Among others, the civil structure may be designed to have separate room for diet preparation, corcera culture, egg production, host culture etc. The host culture room for NPV production should be kept at a distance with proper hygiene and entry may be restricted in such a way to prevent any contamination.

Plant and machinery: There is no requirement of heavy plant and machinery. Racks, trays and other facilities are required for rearing insects. Apart from this centrifuge, mixers and some fabricated equipments for insect collection and rearing are required. For production of *Trichoderma* fermentors, laminar flow apparatus etc. are required. All the machinery required are locally manufactured and available.

Raw material: For rearing of insect special diet is required which comprises of pulses, vitamins, antibiotics etc. For production of *Trichoderma* molasses-yeast medium, is required. All these materials are available locally.

Water: The water requirement is mainly for feed preparation, washing, cleaning, drinking etc.. Water quality should be tested to establish the suitability.

Power: Power supply is essential for bio-pesticide units.

Manpower: Production of bio-pesticides required skilled manpower. There is need for a number of labours at each stage of production. The manpower requirement is *Technical staff* (2-3 nos.), skilled labour (3-5 nos.) and semi-skilled labour (5-10 nos).

1.2 Scale of production

These biopesticides can be produced on a small or large scale. Small scale production is particularly suitable to village or community level cooperatives, which can produce and distribute these for local use. As the production technology of some of these agents (particularly *Trichogramma*) is relatively simple, the local farmers/SHGs can be trained to undertake the production. Medium and large scale production can be undertaken by firms, sugar mills cooperatives engaged in the manufacture and distribution of agro-chemicals. Similarly, seed companies are particularly well placed for undertaking the production and marketing of *Trichoderma*.

Market potential: Considering the negative effects of indiscriminate case of pesticides, importance for organic farming and promotion of sustainable farming practices it is estimated that there are huge scope for new biopesticide units, particularly in the states of Maharashtra, Gujarat, Rajasthan, Madhya Pradesh, Tamil Nadu, Andhra Pradesh, Uttar Pradesh, West Bengal, Karnataka, Uttaranchal, Jharkhand, Kerela, Assam, Jammu & Kashmir, Arunachal Pradesh, Mizoram, Nagaland, Manipur, Meghalaya, Sikkim and Chhatisgarh. In these states organic crops such as pulses, cereals, spices, vegetables and fruit crops are being grown organically.

Regulatory measures: As the bio-control agents are living organisms, it is very important to have effective regulatory measures. Directorate of Plant Protection Quarantine and Storage, Department of Agriculture and Cooperation, Ministry of Agriculture, Government of India has issued guidelines/data requirements for registration of bio-pesticides in the country. As per this, all the units have to meet the Indian standards and technical specifications to be eligible for registration under the Insecticides Act, 1968.

Bio-pesticides registration: At present, *Bacillus thuringensis*, neem based formulations, microbial pesticides like fungi, NPV etc., are included in the schedule of Insecticides Act, 1968. This ensures the quality of bio-pesticides at farmer's level. The standard parameters, protocols for data generation, guidelines for registration are prepared and circulated to prospective entrepreneurs by Ministry of Agriculture, Government of India. Now as such, any person dealing with biopesticides without registration is illegal.

1.3 Bioagents for commercial production

There are different types of bioagents which can be commercially mass produced for large scale distribution among the farmers for control of insect pests. They are:

Trichogramma chilonis, T.brasiliensis and *T. pretiosum* (egg parasites) for tomato fruit borer *Trichogramma chilonis* - for brinjal shoot and fruit borer, shoot borers of cotton, sugarcane, rice etc.	*Cryptolaemus montrouzieri* (Australian ladybird beetle) for control of several species of mealy bugs and soft scales *Chrysopa spp.* (green lacewing bug) - for the control of aphids, white flies etc.	Virus: Nuclear Polyhedrosis Virus (NPV) - for major polyphagous pest like *Helicoverpa armigera* (gram pod borer) and *Spodoptera litura* (Tobacco caterpillar) Bacteria: *Bacillus thuringiences* (Bt) - for control of lepidopterous pests Fungi: *Trichoderma viride* and *Trichoderma harziarum* against soil borne fungal diseases Namatodes: for control of soil-borne grubs, lepidopterans and some foliar pests

Field efficacy of biopesticides: Field efficacy trials have been conducted by State Agricultural Universities and ICAR Research Institutes/Stations to know the extent of pest control by application of biopesticides. The percentage of pest control achieved for selected bio-control agents are as under:

Bio-agent	Efficacy of pest control
Trichogramma spp.	60-90%
Cryptolaemous montrouzieri	100%
NPV	70-80%
Trichoderma viridae	60-90%

Characteristics of effective bio-control agents: While using natural enemies, it is important to have fast growing bio-control organism in the fields which can eventually make the conditions unfavourable for the pathogens proliferation. They should have high mobility to prevent pathogen to develop resistant structures; longevity enough to protect plant during its vulnerable period; and environmental tolerance to sustain activity under different soil and climatic conditions. Mode of action varies from pathogen to pathogen, physical contact, chemical nature of killing component, etc.

1.4 Biopesticide production process

The technical details for production of selected bioagents are given below:

Bio-agent	Production Process in brief	Remarks
Trichogramma spp. (egg parasite)	Mass multiplied by using stored grain pest as a host. The production involves the multiplication of host insect on sorghum grains, allowed to be parasitized by *Trichogramma*. Then egg are clued in cards as "Tricho cards".	Used for control of sugarcane early shoot borer, bollworms of cotton, sorghum stem borer, etc.
Crysoperla carnea (Chrysopid predetor)	Mass multiplied in laboratory on the eggs of stored grain pest.	Controls larval pests in pulses, vegetables / fruits
Cryptolaemus montrouzieri (Ladybird beetle)	Mass multiplied on already mass multiplied mealy bugs with the help of pumpkin as under laboratory conditions.	To control mealy bugs especially on fruits.
NPV of *Helocoverpa armigera* & *Spodoptera litura*	The production starts with raising of pod borer and tobacco caterpillar larvae (host culture) on semi-synthetic diet. NP Virus is smeared on cultured larvae. Then the diseased larvae are collected to obtain virus suspension after blending, filtration and centrifugation.	Use against boll worms in cotton and pod borers.

Bio-agent	Production Process in brief	Remarks
Trichoderma Fungal *spp.*	Multiplied in laboratory and formulated in powder form with the help of carrier material (talc powder).	To control root rot and wilt diseases especially on pulses.
Pheromone lures for *Helicoverpa armigera* & *Spodoptera litura*	Sex pheromones are filled into plastic lures at required concentration with the help of micro pippets and placed into rubber septa. The septa is fixed to the trap.	To trap reproductive males of gram pod borer

The technology described above is indigenous and the scientific aspects of production have been standardised by ICAR Research Institutes and State Agricultural Universities. Machineries and laboratory equipments are available from various manufacturers and are of BIS standards.

2. Biocontrol Units for Insect-Pests

2.1 Production technology of *Trichogramma* egg parasite

Trichogramma spp. belongs to the category of egg parasitoid of biological agents. *Trichogramma spp.*, the most widely used bio-control agent and is effective against bollworms of cotton, stem borers of sugarcane, fruit borers of fruits and vegetables. It attacks the pest at the egg stage itself and hence damage done by larvae is avoided. It offers a lower cost but more effective plant protection option in comparison to insecticides. Two species i.e., *T. chilonis* and *T. japonicum* are predominantly used in India.

Mode of action: *Trichogramma* are dark coloured tiny wasps and the female wasp lays 20-40 eggs into the host's eggs. The entire cycle is completed within 8-12 days. The tiny adult wasps search for the host (pest) eggs in the field and lay their eggs into the eggs of the pests. The parasitised host's eggs turn uniformly black in 3-4 days. The *Trichogramma* eggs on hatching, feed the embronic contents of host's egg, completes its development and adult comes out of the host egg by chewing a circular hole. A single *Trichogramma,* while multiplying itself, can thus destroy over 100 eggs of the pest.

Equipment needed: Equipments like semi-automatic corcera rearing cages, trays, iron racks, hot air oven, air conditioner, UV chamber, incubator, moth breeding tins, grinder, mating chambers, parasitization jars, refrigerator, wire mesh, netlon etc. are required for mass rearing of corcera and Trichogramma production. The flow chart for production of *Trichogramma* is given below:

Preparation of sorghum feed material

⇓

Corcyra charging (host insect)

⇓

Collection of moths

⇓

Transfer of moths to oviposition cages

⇓

Collection of eggs

⇓

Cleaning, sieving of eggs & spreading on gummed cards

⇓

Exposure of cards with eggs to UV radiation

⇓

Introduction of Trichogramma spp. (Ratio - 1 : 5)

⇓

Transportation & field release of Trichocards

Flow Chart for mass production of *Trichogramma* spp.

Identification of host: The *Trichogramma* multiplication starts with identification of a suitable host species. In India *Corcera cephalonica,* a stored grain pest has been used for mass multiplication of targeted species.

Rearing of host insect: The host rearing containers are made of materials which are non-toxic, cheap and optimum sized to permit mating and host searching and amenable to easy cleaning. Most commonly used cages are wooden cages, which are now replaced with semi automatic corcyra rearing cages. The nuclear culture, i.e. eggs of *Corcyra cephalonica* is introduced in rearing cages. Semi-automatic rearing cages is most commonly used cage.

Preparation of feed material: Corcyra feed may be prepared from bold white sorghum grains without any insecticide residues. This can be tested by taking a sample of 100 g from each bag. The crushed sample is fed to 20 number of 1st/2nd instar *Corcyra* larvae for 2-3 days. Based on the mortality of the larvae, suitability of grains may be decided. The requisite quantum of sorghum is milled to make 3-4 pieces of each grain. Sorghum grains are heat sterilised in oven at 100^{0}C for 30 minutes and the grains are sprayed with 0.1% formalin. This treatment helps in preventing the growth of moulds as well as to increase the grain moisture to the optimum (15-16%), which used to lost due to heat sterilisation. Then grains are air-dried.

Corcyra charging: In each rearing cage, 7.5 kg of sorghum grains are filled and charged with 0.5 cc eggs (1 cc = 20,000 eggs) of mother culture. Yeast, groundnut kernel and streptomycin are added to enhance egg laying capacity of the adult moths and for enriching the diet.

Collection of moths: After about 40 days of charging, moths start emerging and the emergence continues for two months. About 10 to 75 moths emerge daily with the peak emergence being between 65th and 75th day. Collect the moths daily and transfer to the specially designed oviposition cages for egg laying. Roughly 2000-3000 pairs of moth can be placed in one chamber. Moth emergence reduces after 100 days of initial infestation and cages are released for cleaning.

Collection of eggs: Eggs are collected by means of manual suction and are placed in tubes and counted with measuring cylinder. Approximately 1 cc of eggs of *Corcyra* counts about 20,000 at the fresh harvest. After that due to shrinkage of eggs the count may be increased. The final output of *Corcyra* eggs from one cage is approximately 7.5 cc.

Egg preparation: The eggs of *Corcyra* thus collected are cleaned to make it free from insect scales etc. They are sieved thrice and then poured on a plain paper. By slowly tapping eggs come downward stick on to gummed card. Thus, the cleaned eggs are spread on the gummed cards (15 cm x 10 cm) with the help of screen. These eggs of *Corcyra* are exposed to UV rays of 15 watt UV tube for 45 minutes to prevent hatching. While UV exposure, egg card should be kept about 12-15 cm away from tube.

Introduction of Trichogramma : After the sterilisation the egg cards are placed in plastic bottles and are introduced with nucleus culture of *Trichogramma* species of egg or pupal stage. The ratio of host egg and parasite adult should be maintained at 1:5.

Production of Trichocards: The parasitisation of *Trichogramma* spp., in laboratory condition on 1 cc eggs of *Corcyra cephalonica,* which are uniformly spread and pasted on a card measuring 15 cm x 10 cm is called as Tricho card. The card has 12 demarcations (stamps). About 12,000 Trichogramma adults emerge out from this card in 7-8 days after parasitisation. To delay the emergence of Trichogramma, these cards can be stored in refrigerator at 5-10^{0}C for 10-15 days. On removing the cards to room temperature, the parasitoids emerge normally. Trichocards have a shelf life of 2-3 days. However, these can be stored in a refrigerator for a period of 1 month without any spoilage.

The demand for Trichocards will start from the onset of *kharif* season and extends to *rabi* season. The summer season vegetables offer an extra demand.

Use of Trichocards: The cards are to be used before the emergence of the adult parasite. Cut or tear each Trichocard into small pieces and distribute them all over the field. The pieces may be stapled to plant leaf at 7-8 m distance. Care is to be taken to release the parasites either in morning or evening i.e., during cool hours, in windward direction and there should not be any pesticide spray. Before releasing the parasite, the infected shoots are to be cut to ground level and buried inside the soil so as to avoid secondary infestation.

Trichocards should be packed in such a way that the parasitised surface is on the inner side. Emergence date should be specified on cards for the guidance of the users. Trichocards should be stapled on the inner-side of the leaf to avoid direct sunlight. Card should be stapled in morning hours and just before emergence to avoid predation. Farmers should refrain from using pesticides in the field where *Trichogramma* are released.

Field release and application: For controlling sugarcane early shoot borer, release 6,000 parasites per week per acre area, for a period of 5 weeks, starting from 4th week of planting i.e., as soon as the adult male moths of early shoot borer are noticed in the field. Totally 30,000 parasites are to be released per acre. More parasites may also be released depending upon the crop and pest density. In cotton, the Trichocards are released in the field at 45 days after sowing @ 5 cards/ha (one lakh eggs). At least three releases are necessary.

2.2 Production technology of Chrysopid predetors

Chrysopid predators are important for the management of bollworms and several sucking pests in fruit crops and aphids in cotton. They are capable of bringing down the population of the pest drastically. Chrysoperla *(Chrysoperla carnea)* is a potential chrysopid, which is also amenable to mass multiplication.

Chrysoperla are generally green in colour, varying in length from 1.0-1.3 cm. The pre-oviposition period lasts 3 to 7 days. Adults start laying eggs from 5th day onwards and peak egg-laying period is between 9 and 23 days after emergence. The male longevity is 30-35 days. On an average an adult female lay eggs of 600-800 eggs/female . The eggs are stalked and green in colour. The eggs are laid singly or in clusters. Egg stage lasts 3-4 days. The larva has 3 instars and after 8-10 days it will form cocoons. Adult emerges in 5-7 days from cocoons.

The green lacewing is being mass released in the field for the control of aphids, white flies, mealy bugs and eggs and young larvae of lepidepteron pests. The *Chrysoperla* predetors may be used on groundnut, pulses,vegetables, cotton, ornamentals and several other crops. They also feed on the eggs and freshly hatched larvae of *Helicoverpa armigera* and such other caterpillar pests.

It is being mass produced primarily on the eggs of rice grain moth, *Corcyra cephalonica* in India. For mass production of chrysoperla, an efficient rearing technique is required.

Equipment required: Facilities like rearing room (6 x 6 m), slotted angle iron racks, work tables, plastic louvers 60 x 22 cms with 2.5 cm cubical cells, acrylic sheets to cover the louvers, glass vials, adult oviposition cages (45 x 30 x 30 cms), plastic louvers, plastic containers, scissors and brushes, cotton wool, tissue paper, sponge, fructose, protinex, honey, yeast, castor pollen etc. are required for the mass rearing of chrysopids.

Production methods: Chrysoperla predators are mass multiplied in laboratory at 27 ± 1^{0}C and 70% RH on the eggs of *Corcyra cephalonica*, a laboratory host. Three days old 120 chrysopid eggs are mixed with 0.75 ml *Corcyra* eggs (the embryo of Corcyra eggs are inactivated by keeping them at 2 feet distance from 30 watt ultraviolet tube light for 45 minutes) in a plastic container. On hatching, the larvae feed on the contents of eggs. The second and subsequent instars are reared individually in cells of louvers on the eggs of *C. cephalonica*. It is assumed that for rearing 100 larvae (1 cc) *C. cephalonica* eggs are required. Host eggs are provided twice during the course of larval rearing. First feeding of 1.75 ml for 100 larvae and second feeding of 2 ml for 100 larvae with a gap of 3 to 4 days is provided. Cocoons formed in the cells are collected after 24 hours. The cocoons are placed in oviposition cage for adult emergence.

In each oviposition box roughly 20 pairs can be accommodated and inside portion of the container is covered with black paper on which adults lay eggs. The adults in the oviposition boxes are provided with castor pollen, protinex mixture (equal volume of protinex, fructose, honey and powdered yeast dissolved in small quantity of water), 50% honey and drinking water in cotton swab. Adults lay eggs on the under surface of the top lid which is removed by sliding a clean lid. After 24 hours of hardening the eggs are gently brushed with a brush to dislodge on to a paper. Eggs are collected and either reused for mass multiplication or sent to farmers for field release. Only first instar larvae are released on to the recommended crop plants.

Field release: At least 1000 eggs or larvae may be used per acre.

2.3 Production technology of Australian ladybird beetle (*Cryptolaemus montrouzieri*)

Mealybugs are serious pests on fruits, vegetables, ornamentals and plantation crops. Besides causing direct loss to the plants they also reduce market value of infested fruits. The extent of damage may go upto 70 percent in severe infestation. Ladybird beetle, *Cryptolaemus montrouzieri* introduced from Australia is a potential bio-control agent and is being utilized on many crops in Southern India.

Mealybugs or scale insects constitute the natural food of certain ladybird beetles. The adult beetles as well as their larvae (grubs) seek the pests and feed voraciously on all stages. They often wipe out the entire pest colonies. The ladybird beetles are being used for suppression of mealy bugs in citrus, coffee, grapes, guava, ornamental and a variety of other crops.

Equipment needed: Equipments like wooden boxes/cages, iron rack, buckets etc. are needed for mass multiplication of ladybird beetles.

Production methods: The production involves the following steps:

- After 15 days of infestation of pumpkins with mealy bugs (*Planococcus citri*), they are exposed to a set of 100 beetles for 24 hrs. After exposing, the pumpkin is kept back in a cage. The beetles during the period of exposure feed on mealybugs as well as deposit their egg singly or in groups of 4-12. The young grubs feed on eggs and small mealybugs but as they grow they become voracious and feed on all stages of mealybugs. For facilitating the pupation of grubs, dried guava leaves or pieces of papers are kept at the base of each of the eggs. The first beetle from the cages starts emerging on 30th day of exposure to beetle adults. The beetles are collected daily and kept in separate cages for about 10-15 days to facilitate completion of mating and pre-oviposition. The beetles are also fed on diet containing agar powder (1 gm), sugar (20 gm), honey (40 cc) and water (100 cc).

- The adult beetle diet is prepared by boiling sugar in 70 cc of water, adding 1 gm agar, diluting 40 cc honey in 30 cc of water and adding to the sugar and agar mixture when it comes to boiling point. The hot liquid diet is kept on small white plastic cards in the form of droplets which get solidified on cooling. Such cards containing diet can be fed not only to *C. montroozieri* but also to many other species of cocinellids. From each cage about 175 beetles are obtained. The emergence of the beetles is completed within 10 days.

The Beetles can also be reared on *corcyra cephalanica* eggs but empty ovisacs of *Planococcus citri* are to be kept for inducing egg laying by the beetles.

Field release and application: Before releasing in the field in the endemic areas, moderate to severely infested plants are marked. Measure should be taken to stop the patrolling of ants on the trunk at least 3 days. Release of 10-15 adults/tree depending up on canopy and infestation once in a season @ 600 to 1000 beetles may be released per acre.

All due precautions should be taken to avoid scarcity of food for the grubs to avoid cannibalism. All the pumpkins showing sign of rotting should be properly incinerated.

2.4 Production technology of NPV

Baculovirus group has a very narrow host range and generally infests the larvae of crop pests. The research aimed at insect pest control is, therefore, confined to nuclear polyhedrosis viruses (NPVs) and granular viruses (GVs). NPV is a nucleic acid (double standard, circular DNA) enclosed in protein matrix, hence it is called polyhedral occlusion body (POB). NPV infects the nucleus of the cell and multiplies within the nucleus.

Nuclear Polyhedrosis viruses like Ha-NPV, Sl-NPV are increasingly being used as alternatives to chemicals. These viruses have distinct advantages over other methods of pest control. These viruses are highly specific and do not affect beneficial insects like parasitoids and predetors and are safe to fish, birds, animals and man. Considering the usefulness of NPV's there has been a growing demand amongst the farmers for these bioagents.

In India, extensive research has been conducted on the use of NPVs for tackling two major pests namely *Helicoverpa armigera* and *Spodoptera litura.*

Equipment required: The major equipments like centrifuge, laminar flow, magnetic shaker, microscopes, autoclave, coolers, refrigerators, incubator, distillation units etc. are required in addition to glassware, plastic trays, basins, iron racks etc. for mass production of Ha-NPV and SI-NPV.

Process of mass production of Ha-NPV and Sl-NPV: The mass production of Ha- NPV and Sl-NPV involves 3 steps:

i. Rearing of adult gram pod borer and tobacco caterpillar for mass production of eggs.

ii. Rearing of larvae of the above species either on the host plants like chickpea and castor under semi natural condition or on the synthetic diet in the laboratory conditions.

Inoculation of Ha-NPV and SI-NPV into the larvae of gram pod borer and tobacco caterpillar respectively for mass multiplication of viruses and extraction of polyhedral occlusion bodies (POBs) from the diseased larvae, are used as biopesticide on the crop plants.

Field collection of eggs/larval stages of host insects from infested fields or adults from light/pheromone traps

⇓

Procurement of nucleus culture of host insects from insectaries or mass rearing units

⇓

Transfer of moths to oviposition cages

⇓

Collection and surface sterilisation of eggs

⇓

Rearing of larvae on natural or semi-synthetic diet

⇓

⇓

Inoculation of larvae with NPV and rearing of diseased larvae

⇓

Collection and surface sterilisation of pupae

⇓

Checking of quality of NPV based on POB counts and containerisation of NPV

⇓

Transfer of pupae to adult emergence cages

⇓

Field application

Flow chart for mass production of Nuclear Polyhedrosis Virus

Diet preparation: The larvae of gram pod borer and tobacco caterpillar can be multiplied by using chickpea based semi-synthetic diet. The composition of the diet for rearing larvae is as follows:

	Item	Quantity
'A' fraction:	Chickpea (Kabuli chenna) flour	105.00 gm
	Methyl para-hydroxt benzoate	2.00 gm
	Sorbic acid	1.00 gm
	Streptomycin sulphate	0.25 gm
	10% formaldehyde solution	2.00 ml
'B' fraction:	Agar-agar	12.75 gm
'C' fraction:	Ascorbic acid	3.25 gm
	Yeast tablets	25 tablets
	Multivitaplex	2 capsules
	Vitamin E	2 capsules
	Distilled water	780.00 ml

About 390 ml of water is mixed with fraction 'A' of the diet in the blender which is run for two minutes. Fraction 'A' and 'C' are mixed and the blender is run again for 1 minute. Fraction 'B' is boiled in the remaining 390 ml water, added to the mixture of A and B and the blender is run for a minute. Formaldehyde solution is added at the end and the blender is again run for a minute.

Production of eggs of tobacco caterpillar: The culture of tobacco caterpillar is initiated by collecting eggs from the fields of castor, cauliflower, lucerne, tobacco etc. These field collected eggs are reared in isolation to eliminate the emerging parasitoids and diseases, if any. The culture can also be established by collecting the gravid females with the help of light traps. Once the pure culture is established the mass production is commenced under laboratory conditions after the first generation established.

Pairs of newly emerged moths of tobacco caterpillar are placed in well ventilated plastic containers. The inner wall of the containers is lined with paper to enable the adults to lay eggs. The bottom of the container is lined with sponge covered over by blotting paper. The moths are provided with 50% honey solution and water on two cottons swabs placed in small plastic cups. The eggs which are generally laid in batches on the paper are cut out. Freshly laid egg masses are sterilised by dipping in 10% formalin for 30 minutes, washed in running water for 30 minutes, dried on blotting paper and kept for hatching in sterilised glass vials.

The freshly laid eggs can also be surface sterilised in 0.05 percent solution of sodium hypochlorite for 5 minutes. These eggs are washed several times in running tap water to remove the traces of sodium hypo chlorite. The traces of sodium hypochlorite could be neutralized by dipping the eggs in 10% sodium thiosulphate solution and again the eggs are washed thoroughly under running tap water. The surface sterilised eggs are kept in plastic tubes (7.5 x 25 cm) on moist tissue paper for continuing the stock culture. After 3 days, the newly hatched larvae are transferred to bouquets of castor leaves and kept in a plastic container with stand for pupation. The pupae are collected 3 days after all the larvae enter the sand. The pupae are sexed and kept on a lid over a wet sponge in adult emergence cage. After 10 days, freshly emerged males and females are collected from their respective emergence cages. Tobacco caterpillar larvae can be multiplied on a chickpea based semi-synthetic diet composition as described above under 'diet preparation' section.

Production of gram pod borer (Helicoverpa armigera): The process is initiated either collecting the adults with the help of light traps. It could be by collection of larvae on a large scale from its host crops in endemic areas. Nucleus culture can also be obtained from the established laboratories. The material thus obtained is reared in the laboratory in aseptic conditions and the healthy progeny is selected and established.

The production starts with the availability of 250 pairs of adults every day, which will yield 10,500 eggs daily. The adults are kept @ 100 pairs in each oviposition cage with a cloth enclosing the frame. A circular plastic mesh (on which cotton swabs soaked in water and honey solution are placed in small containers) rests on a support above the base of the frame. The cloth cover is open at both ends with a 20 cm vertical slit in the centre which can be closed with a zip or cloth clips. The cloth cover enclosing the frame is tied with rubber bands at both ends. It is placed on tray with a sponge at the bottom soaked in water. The temperature inside the cage is maintained at 26^0 C and humidity at 60 - 90%.

The eggs are laid all over the inner surface of the cloth cover. The egg cloth is removed daily. This cloth is surface sterilised in 10% formalin for 10 minutes, the eggs could also be surface sterilised using 0.2% sodium hypchlorite solution for 5-7 minutes and treated with 10% sodium thiosulphate solution to neutralise the effect of sodium hypo chlorite, rinsed in distilled water. The eggs are later placed on paper towell under laminar flow for drying. The dried cloth pieces containing eggs are kept in 2 litre flasks containing moist cotton. Flasks are plugged with cotton wrapped in muslin cloth and the bottom of the flask is wrapped with aluminium foil.

***Rearing of larvae of tobacco caterpillar on semi-synthetic diet*:** For rearing of early instar larvae, a rearing unit is prepared by placing a sponge piece on a glass sheet. The sponge is covered with a single layer of soft tissue paper. A small plastic container containing 200 surface sterilised eggs of tobacco caterpillar is placed in the centre over the tissue paper. A petri dish containing about 200 ml of diet is placed inverted over the tissue paper. The eggs hatch within 25 hr and neonate larvae crawl and spread out on the diet.

Late instar larvae are reared in modified plastic boxes. One window each on the four sides of the box is cut and covered with a fine plastic mesh to provide sufficient ventilation and to prevent moisture accumulation inside the box. A thick layer of sterilised sand is spread at the bottom of the box. A small piece of tissue paper is kept at the centre over the sand.

The diet in the petridish (containing 200 larvae) is divided into five equal pieces. One piece of diet bearing 40 larvae is kept in plastic box over the tissue paper so that the sand does not soil the diet. In this way, 5 boxes are charged with larvae from 1 petridish. A plastic grill is fitted into the box in such a manner so that it forms a crest higher than the brim of the box. Thick cake of diet (about 500 gm) in a petridish is divided into two equal pieces. One such piece is kept on the top of the crest and the lid of the box is then fixed so that the diet and grill crest are opposed to each other just beneath the lid. After consuming the small quantity of diet on tissue paper the larvae crawl and perch on the grill and feed from the ceiling of the box. The boxes are stacked and left intact for 3 days. During this time the diet is almost completely consumed. Now another piece of fresh diet (about 250 gm) is kept on the crest in each box and the boxes are closed and stacked again. During the last 3/4 days of larval stage the food consumption is maximum and so is the fecal matter accumulation on the sand layer. After 20 days from hatching the larvae move into the sand and start pupating. In a period of 25 days, all the larvae, pupate and the chitinisation of pupae is also completed. The boxes are now ready for the pupal harvest. The pupae are collected, cleaned, sterilised and placed in adult emergence cages. The freshly emerged moths are then placed in oviposition cages.

***Rearing of larvae of gram borer on semi-synthetic diet*:** The larvae of gram borer can also be reared on a chickpea based semi-synthetic diet. The diet is poured as per the requirement either on the nylon mesh for rearing 5-7 day old larvae or in tray cells for rearing the older larvae or poured into sterilised petridish and allowed to solidify. The diet could be stored in the refrigerators for up to 2 weeks. For preparing large quantities of diet, the quantity of diet ingredients to be used should be calculated accordingly and industrial type blenders could be used.

The larvae are removed from the top of the aluminium foil wrapped flasks with a brush and then transferred to the diet. About 220 larvae are transferred to diet impregnated on nylon mesh and placed in plastic containers or sterilised glass vials. About 100 such containers are maintained daily for 5-7 days. A multi-cellular tray with semi-synthetic diet is advantageous for rearing a large number of larvae.

Starting with 10,500 eggs, the total number of larvae available is 10,000 considering an estimated 5% mortality in initial 5 days of emerging and 10% mortality up to first 5 - 7 days. The total number of larvae available for virus production is 8000 (80%). The rest of 20% will be utilized for maintenance of host culture continuously.

The diet requirements at various stages of production of larva are:

1. For the young larvae up to 5-7 days will be 2-4 gms/larva.
2. For 5-7 day old larvae for *Ha-NPV production* will be 4gms/ larva.
3. For five to seven days old larvae for *continuation of host culture* will be 6 gms/larvae.
4. For rearing the *field collected larvae* for augmenting the nucleus stock will be about 1 kg.

In host culture units, larvae start pupating when they are 18-19 days old and the pupation will be over within 2-3 days. The harvested pupae are surface sterilised using 0.2% sodium hypochlorite solution followed by washing with 10% sodium thiosulphate solution to neutralize sodium hypochloride and then washed thoroughly with distilled, sterilised water. After washing, the eggs are dried by rolling over blotting paper. The male and female pupae are separated out and placed over moist sponge in adult emergence cages.

The egg, larval, pupal and adult stages of gram borer last 3-4, 18-29, 7-8 and 7-9 days, respectively. The oviposition period of the females is about 5 days.

Production of Helicoverpa armigera NPV (Ha-NPV) and Spodoptera litura NPV (Sl-NPV): For Ha-NPV and Sl-NPV production, the synthetic diet prepared is poured at 4gm/cell in the multi-cavity trays and the diet surface is uniformly sprayed with virus prepared in distilled sterilised water at 18 x 10^6 POBs/ml. Eighty percent of the total 5-7 day old larvae can be utilised for Ha-NPV and Sl-NPV production.

The trays are incubated at 26^0 C for 7 days. In case of virus infected larval trays, the diseased larvae dies after attaining its maximum size of 6th instar, where the dead caterpillar will have 2-6 billion poly occlusion bodies (POB) which is in terms of larval equivalent (LE). *H. armiegera* NPV of 1 LE = 6×10^9 POBs; 1 LE of *S. litura* = 2×10^9 POBs. The dead larvae have to be harvested, macerated in distilled/sterilised water and filtered through muslin cloth to get the crude suspension of the virus. The extraction is centrifuged to further clarify the solution.

General precautions: In production units, keep the host culture in a separate room and the virus production and storage facility should be located in a different facility. Following care should be taken for commercial production of NPV:

- In the NPV production units, inspite of best care, 100% larvae are not infected, the larvae which do not turn inactive after 4 - 5 days and keep consuming the normal diet should be culled out regularly from the NPV production unit.
- Utmost care should be taken to prevent the break in the chain of the production system. This could be achieved only if highly dedicated and disciplined workers are engaged for such production units.
- Strict hygiene should be maintained in different facilities. The equipments used should be either heat sterilised or sterilised using steam or chemicals. The work place should be thoroughly disinfected with sodium hypo chlorite solution.
- The host culture should be initiated from a batch of healthy adults.
- Microbial infection could be avoided if good insect husbandry practices are followed. If infection is detected, the culture or infected part should be destroyed immediately. Besides hygienic conditions, optimum temperature (24°C-26° C) and humidity (65-70%) should also be maintained.
- The texture and quality of the natural/semi-synthetic diet should be good.

- entry to host culture unit after visiting virus production unit should be avoided.

Mode of action: NPV acts as a stomach poison only to the target host (pest) and hence beneficial insects are not affected. The infected larvae become pale and glossy and tissue get disintegrated and liquified. Most of the body tissues and organs (except gut) get infected by polyhedralocclusion bodies (POBs), which contains the virions. The liquid which oozes out of the infected larvae (which hang upside down) contains millions of POBs. Each POB measuring about one micron in diameter and possessing a characteristic movement can be identified under the microscope.

Use of Sl-NPV: The virus is specific and infects only tobacco caterpillar. NPV can be successfully multiplied on tobacco caterpillar and the viral extraction can be applied in the field to control the caterpillar. For continuous production of SI-NPV, it is necessary to rear tobacco caterpillar larvae continuously in a laboratory condition.

Use of gram pod borer (Helicoverpa armigera): It is widely distributed in India and infests/damages a variety of cultivated and wild plants throughout its distribution range. It is a serious pest on commercial crop like cotton, pulses like redgram and bengalgram, vegetables like tomato, bhendi and dolichos bean, oilseeds like sunflower, soybean and safflower and cereals like sorghum and maize.

Use of Ha-NPV: Ha-NPV is a highly infective microbial biopesticide which can be used to control Gram borer. It is derived from naturally diseased or under laboratory conditions artificially infected larvae of gram borer.

Field application and dosage: Ha-NPV is used for controlling *H. armigera* attacking cotton, redgram, bengalgram, tomato, okra, sunflower, groundnut, chillies, maize, sorgram etc., whereas, Sl-NPV is used for controlling tobacco caterpillar attacking tobacco, groundnut, soybean, sunflower, cotton, cabbage, beetroot, cauliflower etc.

Directions for use of NPV: The recommended dosage is 200 ml of NPV/acre or 500 ml/ha containing 100 and 250 larval equivalent (LE) of NPV respectively as active infective material (1LE = 6×10^9 POBs). 100 ml of NPV could be diluted in 200-400 litres of water when high volume sprayer is used and in 50-70 litres of water in case of power sprayers. Preferable to spray using high volume knap-sack sprayer. Virus should be sprayed during evening hours. Spray should be initiated as soon as some newly hatched larvae are observed or three to five days after a trap catch of 5 moths per pheromone trap. Subsequent sprays should be made at 7-10 days intervals depending upon the pest population.

2.5 Sex pheromone traps

Sex pheromones are single or complex blend of different chemicals released by one insect to attract the opposite sex of the same species. In general, females (especially the moths) emit sex attractants to attract males for mating. Sex pheromones are artificially synthesized in the laboratories and supplied as sex pheromone lures. Such pheromones are placed in the field to attract trap and kill the males, thus matting is not allowed.

Sex pheromones help in monitoringand early detection of pests (at moth stage only). No harmful effects to beneficial insects, non-target organism or an environment have been observed. It also helps in scheduling pest control measures. Hence, sex pheromone traps can be considered as a key component in organic farming.

Ready-to-use Sex pheromone lures and traps are available for *Helicoverpa armigera* (attacking crops like cotton, redgram, tomato, okra, sunflower, chillies, maize, sorghum etc.) and *Spodoptera litura* (attacking crops like tobacco, groundnut, sunflower, cotton, cabbage, beetroot, cauliflower, etc.)

Equipment needed: Only micropipettes are required in addition to rubber septas, traps and pouches.

Production of Pheromone traps: Sex pheromones are insect specific, produced artificially in laboratories and they are generally imported. In India, it is available from National Chemical Laboratory (NCL), Pune. Chemicals obtained from laboratory is diluted to the required dosage and filled into plastic lures with the help of micro-pippets and closed with rubber septa. Lures are individually sachet packed and should be stored under refrigerated conditions when not in use.

Field application: Lures containing sex pheromones are placed into insect trap and erected in the field at a recommended spacing. The lure will release the sex pheromone at a constant rate over a period of 2-4 weeks. Male moths are attracted and while attempting for matting, fall into a container having pesticide. Thus the female moths in the field are deprived of successful mates and fail to reproduce or lay viable eggs.

Dosage: Timely use of sex pheromone helps in early detection and prompt action against pests. In general, 2-3 traps/acre are recommended for 'monitoring' or more for 'mass-trapping'. These are arranged such that the trap is 1-2 feet above the crop canopy. On the field each lure is effective for at least 15 days. Change the lures once in two weeks.

3. Biocontrol Units for Plant Disease

Crop losses due to soil borne plant pathogens worldwide are *Pythium spp., Fusarium oxysporum, Sclerotium rolfsii, Rhizoctonia solani and Phytophthora* spp. These fungi pathogens generally cause wilt disease in many crops.

3.1 Production technology of Trichoderma fungi

Trichoderma, a fungi, which grow saprophytically in soils have proved as an effective biocontrol agent of wilt diseases. *Trichoderma* spp. are commonly found in almost any soil and other natural habitats consisting of organic matter such as decaying bark, plant material, etc. They grow towards hyphae of other pathogenic fungi, coil them and degrade their cell walls. This process is called "mycoparasitism", which limits the growth and activity of plant pathogenic fungi. In addition, they produce toxic metabolites which protect the seeds from soil borne pathogenic fungi, by forming a protective coating on them.

Trichoderma spp. are saprophytic fungi that grow best in neutral and acid soils and thrive well in moist conditions. The important species available for mass production are *Trichoderma viride and Trichoderma harzianum*

Equipments required: Equipments like fermentor, rotary mixer, auto packer, rotary shaker, laminar flow, water distillation unit, refrigerator, haemo cytometer etc. are required for the production of *Trichoderma* fungi.

Production of Trichoderma: The pure mother culture of *Trichoderma* fungi is being maintained in Agricultural Universities, IARI, some ICAR institutions (like PDBC, Bangalore) etc. The mother culture can be purchased from the identified sources. They have to be further sub-cultured and maintained purely for mass production by adopting standard techniques under the supervision of trained microbiologist or pathologist.

The culture has to be mass multiplied in two levels namely (i) at primary level using shakers in flasks, and (ii) secondary stage multiplication in fermenters. The important factor in this is the preparation of growing medium in which the culture is mass multiplied. After the growing media (molasis and protein material) is formulated and sterilised in fermenter, *Trichoderma* fungi is inoculated using the culture multiplied in the flasks.

The molasses based culture media is continuously aerated by passing sterile air from compressors. After about 3-4 days fermentation period, the culture will be ready for packing in a carrier material. While the inoculated culture is gathering ready in the fermenters, the carrier material is sterilised in autoclaves and kept ready for mixing the culture. Talk powder is reported to be the commonly used carrier material for *Trichoderma* fungi.

The cultured (fungi) and sterilised carriers are mixed mechanically in a blender and the material is packed using semi automatic packing and sealing machine.

Dosage: Talc based formulations of the fungal antagonists are applied at the rate of 4gm per kg of seed for controlling soil-borne plant diseases. Mix the powder with sufficient quantity of water to make slurry for treating seed before sowing.

Application: *Trichoderma* application is eco-friendly and can be used along with organic manure. Soil application of *Trichoderma* spp. is also known to suppress activity of soil-borne fungal pathogens and plant parasitic nematodes (root-knot nematodes). *Rhizoctania solani* and *Pythium* spp. protects transplanted seedlings by colonizing their roots.

Seed treatment is an alternative approach to introduce *Trichoderma* spp. into the soil. This method requires smaller amounts of biological material than soil treatment. Unlike chemical fungicides, *Trichoderma* spp. provides long-term protection without any adverse side effects.

4. Pesticides Production Unit

4.1 Unit Size

The size of the unit biopesticides production unit is given below:

Bio control agent	Capacity per year
Trichgramma cards	16,200 cards per year
Crysopid grub/larvae	6000 per year
Cryptolaemus beetles	5250 beetles per year
Ha NPV & SINPV	Ha-NPV-18000 bottles
	Si-NPV- 12,000 bottles
Trichoderma fungi	2000 kg
Pheromone lures	1000 lures

4.2 Techno-economic parameters

4.2.1 Trichogramma, Chrysoperla and Cryptolaemus

Following materials are required for production of *Trichogramma, Chrysoperla* and *Cryptolaemus*:

A. Equipments

Sl. No	Item	Quantity
	Trichogramma	
1.	Netlan	3 nos.
2.	Mosqito net (105 meters of 5'x20' size)	105 m
3.	Semi-automatic corcera rearing cages	30 nos.
4.	Iron Rack (2mx1mx45cm with 6 partitions)	1 no.
5.	Hot air oven (for dry sterilization of Sorghum)	1 no.
6.	UV Chamber (for corcera)	2 nos.
7.	UV tube light	2 nos.
8.	BOD incubator	1 no.
9.	Grinder/blender	1 no.
10.	Mating chambers	20 nos.
11.	Parasitization jars	20 nos.
12.	Measuring cylinders	5 nos.
13.	Moth breeding tins	4 nos.
14.	Moth scale-egg separator	1 no.
15.	Refrigerator (165 lit capacity)	2 nos.
16.	Bulb plus wire set for trapping pests like Bracon Hebtor	--
17.	Wiremesh (40 mm) to windows	--
18.	Air Conditioner (A.C)	2 nos.
19.	Exhaust fans	3 nos.
20.	Office furniture & tables & chairs, stools	--
	Chrysoperla	
1	Adult oviposition cages for chrysopa (size 45x30x30cm)	6 nos.
2	Plastic trays (1 cu m) with glass plates (6x6 m)	25 nos.
3	Plastic trays for adult feeding	5 nos.
4	Weighing balance	1no.
5	Slotted angle iron rock	1 no.
6	Miscelleneous items (like scissors, brushes, wash bottles)	--
	Cryptolaemus	
1.	Wooden boxes/cages (size 45X30x30 cm)	5 nos.
2.	Iron rack	1no.
3.	Buckets	2 nos.

-- as per requirement

B. Raw materials and output

Trichogramma	
Capacity	*60 cards per day*
No. of cycles per year	*3*
Duration of one cycle (Corcera)	90 days
Life cycle of Trichogramma	8-12 days
Jowar	7.5 kg/cage/cycle
Yeast	1.35 kg/cycle
Groundnut kernel	27 kg/cycle
Streptomycin	1 kg/cycle
Inoculation of mother culture	0.12 cc
Temperature required	27-28^{0} C
RH	80-85%
No. of corcera cages required	30
Corcera sex ratio	1:1
Fecundity	300-350 eggs
Output of corcera eggs	7.5 cc per cage
Yield of corcera eggs/tray	15
No of cards per month	1500
Corcera eggs per cc per year	20000
No. of parasitoids per card	12000
Chrysopid	
Installed Capacity (No. of Larvae)	6000 per year
Capacity utilization (% of production)	0.6
Life cycle (for pupation-2 weeks and for adult emergence-1 week)	3 weeks
1 cc corcera eggs	100 larvae
No of eggs per box	6000
Production (No of Larvae)	3600
Cryolaemus montrouzieri (Lady bird beetle)	
Installed Capacity	5250
No of cycles per year	6 cycles/Larvae
Pumpkin requirement (@ 5 per box, 6 boxes for 6 cycles)	150
Production (No of Larvae)	3150

C. Production Capacity

S.No	Particulars	1st year onwards
A	*Tricho cards*	
	Installed Capacity @ 50 cards per day	16200
	Capacity utilization	0.6
	Production (No of cards)	9720
B	*Chrysopid*	
	Capacity (No of Larvae)	6000
	Capacity utilization	0.6
	Production (No of Larvae)	3600
C	*Lady Bird Beetles*	
	Installed Capacity	5250
	.Capacity utilization	0.6
	Production (No of Larvae)	3150

D. Economics of the unit

Sl. No	Particulars	1st year onwards
	Income (Rs.)	
A	Sale of Tricho cards (@Rs.50 per card)	349920
B	Sale of Chrysopid larvae	2430
C	Sale of Lady Bird Beetles (@Rs.0.75/beetle)	2126
	Total Income (A+B+C)	*354476*
II	Expenditure	
1	Establishments*	154260
2	Cost for production of *Trichgramma*	31254
3	Cost of production of *Chrysopid*	1395
4	Cost of production of Ladybird Beetle	3645
5	Selling expenses @ 5%	17723
	Total expenditure	*208277*
III	Gross Income	146198

* one time expenditure (1st year)

4.2.2 Ha-NPV and SI-NPV, Trichoderma and pheromone traps

Following materials are required for production of Ha-NPV and Sl-NPV, *Trichoderma* and pheromone traps:

A. Equipments

Sl. No	Item of investment	Quantity
1	Laboratory Buildings - 1200 sq ft	2 nos
2	Poly house for host production	1 no.
3	Air Conditioner	2 nos
4	Magnetic shaker and Rotary shaker	1 no. each
5	Phase contrast microscope	1 no.
6	Centrifuge	1 no.
7	Laminar flow chamber (15 cubic feet)	2 nos.
8	Autoclave (Horizontal, 10 lit capacity)	1 no.
9	Stereo Binocular	1 no.
10	Air cooler (300 lit capacity)	1 no.
11	Refrigerators	4 nos.
12	Incubator (10 kg capacity)	1 no.
13	Grinders	4 nos.
14	Electronic balances	2 nos.
15	Plastic trays (60 for Ha-NPV and 40 for Sl-NPV)	2500 nos.
16	Iron racks (12 trays/rack)	55 nos.
17	Plastic basins	350 nos.
18	Tubes	50000 nos.
19	Distillation units (capacity 6-8 lit/hr.)	2 nos.
20	Haemo cytometer for CFU test	2 nos.
21	Plastic containers, oviposition cages, plastic boxes	--
22	Compound microscope	1 no.
23	Micropipette for pheramone traps	--
24	Stabilizers	3 nos.
25	Oven	1 no.
26	Fermentor	1 no.
27	Auto packer/sealing machine	1 no.
28	pH meter (digital)	1 no.
29	250 ml containers for packing NPV suspension	400 no.
30	Steel trolley	2 nos.
31	Glassware	3 nos.
31	Gas connections	2 nos.
32	Office furniture, computer and accessories	--
33	Vehicle	1 no.

-- as per requirement

B. Production capacity

	NPV	
1	Installed capacity (no. of bottles)	
	Ha-NPV	18000
	Sl-NPV	12000
2	Production (no. of bottles)	
	Ha-NPV	9000
	Sl-NPV	6000
	Trichoderma Fungi	
1	Installed Capacity (in kgs) @200 kg per month	2000
2	Production (in kgs)	1000
3	Requirement of carrier material (kg/month)	200
4	Cost of carrier material (Talc powder (Rs./kg)	4
Pheromone Traps		
1	Installed Capacity (in nos.)	
	Helicoverpa traps and *Spodoptera* traps	5000 each
2	Production (in nos.)	
	Helicoverpa traps and *Spodoptera* traps	2500 each

C. Economics of the unit

Sl. No	Particulars	1st year onwards
I	Income	
A	Ha-NPV (@8550 bottles/day)	1923750
	Sl-NPV (@5700 bottles/day)	627000
B	*Trichoderma (@900 kg/yr)*	72000
C	*Helicoverpa* traps (@ 2375 nos.)	14250
	Spodoptera traps (@ 2375 nos.)	23750
	Total Income (A+B+C)	2660750
II	Expenditure	
1	Establishments*	1010850
2	Cost for production of NPV	482500
3	Cost of production of *Trichoderma*	159000
4	Cost of production of Pheramone traps	16737
5	Selling expenses @ 5% net sale	133037
	Total expenditure	1802125
	Gross Income	858625

* one time expenditure (1st year)

The projects on manufacturing biopesticide products would be considered for refinance support by National Bank. Therefore, all participating banks may consider financing this activity subject to their technical feasibility, financial viability and bankability.

5. Bordeaux Mixture

Method of Preparation

- Dissolve 1 kg of copper sulphate in 10 litres of water.
- In another vessel, slake 1 kg of quicklime by adding small quantity of water preferably warm water. When slaking is over, add 5 litres of water and stir well to get a uniform suspension of lime.
- Transfer the lime suspension through a sieve to a vessel containing 85 litres of water and stir well. A small quantity of lime solution may be kept separately.
- Add 10 litres of the copper sulphate solution to the 90 litres of lime solution with constant stirring.
- To test the correctness of the mixture, dip a brightened iron knife for a minute in the mixture. If the knife remains bright, the mixture is correctly prepared. If the knife turns rusty brown or if its brightness is lost, add more lime suspension.
- Correctly prepared Bordeaux mixture will turn red litmus to blue and turmeric powder to orange red in colour.

Following care should be taken while making the Bordeaux mixture:

- For dissolving copper sulphate, use copper, wooden or earthenware or plastic pots or drums.
- Use fresh quicklime.
- Bordeaux mixture should be passed through a sieve before transferring to the sprayers.
- Spraying of Bordeaux mixture should be done on the same day of preparation.

6. Non-Pesticidal Preparation

6.1 Neem seed kernel suspension

Method of preparation

- Take 5 kg of dried neem seed kernels and soak them in small quantity of water.
- Grind neem seed kernel and keep the paste in a cloth bag and soak it for 2 hrs.
- Take a big container; fill it with 10 litres of water by measuring with a standard vessel.
- Keep the bag with neem seed kernel paste in vessel, squeeze it thoroughly for 20 minutes while holding in water.
- Filter the milky white suspension using thin cloth and add 100 gms of soap to it.
- Add water to this suspension to make 100 litres of spray solution. This gives NSKE 5%.
- Spray the suspension on the crop, by shaking the sprayer.

Insects control: *Spodoptera, Helicoverpa,* Semiloopers, leaf folders, all defoliators and sucking pests, including mites. To store neem in large quantities add sulphur 1:10@ 0.5 kg per quintal.

6.2 Chilli-garlic extract

Method of preparation

- Take 3 gram of chillies, remove the pedicel and grind thoroughly.
- Soak the chilli paste in 10 litres of water overnight.
- Take 500 g of garlic, grind thoroughly and soak overnight in 250 ml of kerosene.
- Prepare two extracts separately by filtering through thin cloth.
- Prepare third solution by dissolving soft soap/potassium soap @ 75g in one litre.

- Mix the three solution in a container and keep it for 4 hours.
- Filter the mixture using a cloth and dilute to 80 litres and spray.

Insects control: *Helicoverpa, Spodoptera*

6.3 Cattle-dung and urine Extract

Method of preparation

- Take cattle-dung 5 kg and 5 litres of urine and mix them in 5 litres of water.
- Ferment the solution for 4 days by keeping a lid over the container.
- After 4 days, filter the solution and add 100 grams of lime.
- Dilute the solution in 80 litres of water and spray in one acre.
- Spraying cow dung urine solution prevents egg laying by the moth, eg. *Heiothis, Spodoptera* , etc.
- It gives protection against some diseases. Crop looks green and healthy.

7. Technical Consultancy

Setting up of unit for mass production of bioagents especially those which are based on fungi, bacteria and virus is highly technical in nature. The skill required to handle the mass production process is also higher. For scientific and successful setting up of a unit, the entrepreneurs can take consultancy services from the following agencies:

- Project Directorate, Biological control, ICAR, Bangalore
- Indian Institute of Horticulture Research, Hessaragatta, Bangalore
- Central Integrated Pest Management Centre (CIPMC), White field, Bangalore
- Central Institute for Cotton Research (CICR), Nagpur

Chapter 8

Nursery Development

Species and varieties cultivated in organic agriculture systems are selected for adaptability to the local soil and climatic conditions and tolerance to pests and diseases. All seeds and plant material should be certified organic. Plant varieties should be selected to maintain genetic diversity. Organically grown varieties, and varieties known to be suited to organic cultivation should be preferred.

Organic varieties are obtained by an organic plant breeding program. Plant breeders should use breeding methods that are suitable for organic farming. All multiplication practices should be carried out under certified organic management practices. Breeding methods and materials should minimize depletion of natural resources. All multiplication practices except meristem culture shall be under certified organic management. To be an organic variety, only suitable methods of breeding shall be used as listed below:

Variation induction techniques	Maintenance and multiplication
- combination breeding	• generative propagation
- crossing varieties	• vegetative propagation
- bridge crossing	- partitioned tubers
- backcrossing	- scales, husks, partitioned
- hybrids with fertile F1	- bulbs, brood bulbs, bulbils
- temperature treating	- offset bulbs etc.
- grafting style	- layer, cut and graft shoots
- cutting style	- rhizomes
- untreated mentor pollen	• meristem culture

Organic seed and plant materials shall be propagated under organic management for at least one generation, in the case of annuals, and for perennials, two growing periods, or 12 months, whichever is the longer, before being certified as organic seed and plant material.

1. Nursery Technique

Area and location: An area of 0.25 ha. has been considered for a viable nursery wherein 1.25 lakh seedlings can be raised. The size of the nursery may be increased according to the borrowers category, capacity and demand for planting material. The nursery should be on a gently sloping land to ensure proper drainage. It should have water as a perennial source to ensure adequate supply in hot weather and to reduce costs.

Land and polybed preparation: Land preparation of nursery should be done by ploughing and hoeing the land. Initially the nursery should be raised in mother beds and pricked out in polypots. The shape of plot should be rectangular with size of 100m x 25 m. Ten seed polybeds can be raised of 10m x 1m i.e. 10 sq m. The number of polybeds required at this stage is at 1: 12 ratio i.e. 12 polybeds for each of the primary/seed polybeds. The 1.20 lakh seedlings can be raised in a total of 120 polybeds and remaining 5000 of naked rooted seedlings.

Selection of species: The organic nurseries should plan to produce healthy plants covering, fruits, spices, medicinal-aromatic, vegetables, N-fixing tree/shrubs, fodder, non-wood forest produce and even ornamental species having good demand in the locality. Besides this the prevailing agro-climatic conditions in the area should also be taken into consideration while selecting the species.

Hardening of seedlings: The seedlings can be hardened in the nursery by reducing the water supply over a period of time and exposing them to sunlight over different durations. This would make them capable of facing adverse weather conditions once they are transplanted onto the field. The nurseries should be temporary in nature and are of a five-year duration. During the summer months, shading may be provided by using polythene sheets or shading nets. Bamboo mats can also be used for providing shade. Protection measures may be taken like fencing the area with barbed wire.

1.1 Cost of cultivation

The total cost of raising a nursery with 1.25 lakh seedlings has been estimated at Rs.3.14 lakh for the first year. The detailed item wise unit cost is furnished below:

Particulars of works	Unit	Cost. (Rs.)
Site Preparation	8 MD	800
Fencing with Barbed wire for 150 RMT	Rs.50/RMT	7500
Preparation of Compost pit, nursery path	10 MD	1000
Maintenance of Irrigation source	LS	2000
5 HP diesel Pumpset	LS	25000
Cost of Pipeline for irrigation (100 mts.)	Rs.15/RMT	1500
Cost of Implements for nursery operations	LS	2500
Cost of Water Tank	LS	5000
Preparation of Polybeds (120)	100 MD	10000
Cost of Net for providing shade and installation	LS	30000
Preparation of Seed beds (10)	10MD	1000
Cost of Seeds/mother plants	LS	50000
Cost of Polybags (400 Polybags/kg)	Rs.40/kg	12000
Cost of Pot mixture	Rs.120/MT	30000
Cost of Manure/Vermicompost	Rs.10/kg	12000
Cost of Biopesticides for plant protection	LS	2500
Cost of diesel and lubricants for pumpsets	150 ltr.	5250
Cost of thatching material	LS	1000
Cost of sowing on seed beds	10 MD	1000
Cost of weeding and hoeing	50 MD	5000
Cost of picking up from germi beds	50 MD	5000
Filling up of Polybags @ 200 polybags /MD	625 MD	62500
Shifting of polybags	50 MD	5000
Cost of labour for irrigation	100 MD	10000
Cost of organic fertilizer application	25 MD	2500
Cost of application of bio-insecticides	25 MD	2500
Maintenance of paths	10 MD	1000
Maintenance of pumpset	LS	2500
Watch and ward	Rs.1500/ month	18000
Grand Total		314050

2. Plant Propagation by Tissue Culture

The success of quality plant material development depends on selection of desired types of plants and their multiplications. Selection of desired types is based on evaluation of the quantitative and qualitative performance of plants. Over the years, the scientists have developed various techniques for selection of desired types of plants and their multiplication. Recently, interesting developments have taken place in the field of plant multiplication which involves culture of cells or tissues in laboratory.

Traditionally, plants are multiplied by means of seeds (sexual propagation) or organs other than seeds (asexual or vegetative propagation). These organs are usually stems, leaves or roots. Though multiplication by seeds is the cheapest method, it suffers from certain disadvantages. Plants raised from seeds may not repeat good performance of mother plants. Many horticultural plants take a long time to produce seeds/fruits and many of them do not produce viable seeds or desired quality of seeds. Plants propagated vegetatively do not suffer from these disadvantages. However, vegetative propagation is rather a slow, time and space consuming process. Besides, it is usually infected with latent diseases. Some plants are also not amenable to vegetative method of propagation, for example, coconut, papaya, oil palm, clove etc. Plant propagation by tissue culture method, which could overcome disadvantages of propagation by seeds or vegetative organs, was found commercially successful. The method (also known as micro-propagation) involves the culture of whole organism from cells or tissues or plant parts in glass (in vitro) on a defined medium under germ free conditions (sterile or aseptic), whereas conventional method of vegetative propagation (macro-propagation) involves culture of parts into whole organisms in natural conditions.

A large number of true to the type plants can be propagated within a short time and space and that too throughout the year. For example, it may be possible to propagate 2-4 lakhs of *Tissue Cultured Plants (TCP)* from a single bush of rose against 10 to 15 plants by vegetative means. Also, it may take about 2-4 months to produce healthy planting materials by tissue culture means, whereas a minimum of 6-8 months is required for most species by the latest method of vegetative propagation.

- Tissue culture could be a useful way of circumventing or eliminating disease which can accrue in stock plants.
- TCPs may have increased branching and flowering, greater vigour and higher yield, mainly due to the possibility of elimination of diseases.
- The method may succeed to propagate plants where seeds or vegetative propagation is not possible or difficult or undesirable.
- The method saves space and energy. For example, 2500 m^2 of heated green house can be replaced by a climatised room of $10m^2$.
- The flexibility of nurseries can be improved. As the capital investment on mother plants is reduced to almost zero, it may be easier to adapt to changing conditions. Additionally, a better programming of the production is possible, because of the greater plant uniformity and the availability in the mass at any time.
- Tissue culture can be utilised for breeding new varieties, preservation of germplasm and in vitro synthesis of metabolites.

Propagation by tissue-culture offers good commercial prospect in organic fruit plants, vegetables and also ornamental plants, where value of the products is high. The technique has reportedly been successful in more than 100 species of plants. In India, the working group appointed by the Ministry of Commerce has proposed an export target of Rs. 30 crores i.e. about 60 million TCPs over a five year period as against the present production of 8-10 million TCPs. It may not be possible to meet this requirement by conventional nurseries. It would, therefore, be desirable to encourage commercial tissue culture laboratory to supplement the production of planting material by conventional means. Tissue culture method of propagation is highly labour intensive, 55-60% of the cost is on account of labour. India's potential for export of tissue culture organic plants is rated very high because of abundant and cheap labour.

2.1 Requirements for tissue culture plant propagation

In line with the technology and objective of tissue cultural propagation, various facilities are required as indicated below:

Land: It is required to set up laboratory, mother plant unit, green house and office. Space may also be required for installing tube well/dug well and parking of vehicles.

Source of technology: The propagation by tissue culture is much more sophisticated than other types of plant propagation. A tie-up with reputed laboratories could be helpful. However, if well-qualified and experienced staffs are recruited, it may be possible to set up such units without any tie-up.

Mother plants: Mother Plants would serve as source of tissues (explant). Their performance should be tested before use as source of explant. In case of tie-up with well established laboratories, explants from tested mother plants could be available free of cost. Otherwise, collection, maintenance and testing of superior mother plants would be necessary. Plant species reported to respond to propagation by tissue culture are listed below:

Vegetables: Onion, Asparagus, Brassica, Tomato, Chicory, Capsicum etc.

Fruit plants: Banana, Strawberry, Apple, Pear, Cherry, Citrus, Gooseberry, Grapes, Papaya, Pineapple, etc.

Spices and Plantation crops: Ginger, Turmeric, Vanilla, Coffee, Cardamom, Oil Palm etc.

Forest species: Poplar, Eucalyptus, Bamboo, Pinus, Cupressus, Thuja, Sequoia, Ulmus, Spiraea, Betula, Salix, Ilex, Fagus, Picea etc.

Medicinal and Ornamental plants: Philodendron, Gladiolus, Diffenbachia, Lily, Monstera, Kalanchoe, Maranta, Pentunia, Alocasia, Narcissus, *Aloe vera,* Rose, Fern, Mimosa, Gypsophila, Aster, Aglaonema, Hydrangea, Amarylis, *F. elastica, F. lyrata, F. compacta, F. foliole, Ficus robusta, Ficus benjamina, Ficus mini,* Nerine, Tulip, Iris, Nephrolepis, Freesia, Hyacinth, Heliconia, Anemone, Syngonium, Begonia, Dracaena, Eucharis, Peperomia, Caladium, Euphorbia, Carnation, Pelargonium, Chrysanthemum, Platycerium, Gerbera, Pteris, Saintpaulia, Davallia, Streptocarpus, Osmunda, Anthurium, Mammillaria, Aechmea, Christmas cacti, Cyclamen, Easter cacti, Kalanchoe, Agapanthus, Episcia, Asparagus, Spathiphyllum, Gloxinia, Alstroemeria, Guzmania, Hamamelis, Cordyline, Hemerocallis, Schefflera, Dendrobium, Liriope, Cymbidium, Strelitzia, Cattleya, Nandina, Odontoglossum, Rhododendron, Phalaenopsis, Bougainvillea, Vanda, Buddelia, Weigela, Magnolia, Ribes, Deutzia, Epidendrum, Crocsmia, Forsythia, Calathea, etc.

Laboratory: A tissue-culture laboratory generally comprises of media preparation room, media store room, inoculation room, growth room, culture transfer room, sterilization area, washing area, etc. The floor plan should be designed to promote maximum efficiency. The design should facilitate maintenance of optimum temperature, humidity, illumination and ventilation. Inside air of laboratory should be free from dust particles.

Sl. No	Particulars	Area (sq. ft.)
1	Boundary Wall	480
2	Laboratory	5000
3	Auxiliary Structure	500
4	Polyhouse	
	(i) Mother plant area	2000
	(ii) Hardening area for Plantlets	3000
	(iii) Polytunnels (25 nos.)	40 (each)
	(iv) Wirenet for Polyhouse walls	3600
	(v) Sunshade net	2000

Sl. No	Particulars	Area (sq. ft.)
5.	Clean Area	
	1. Media Store and Production Control	200
	2. Post Autoclave Area	150
	3. Culture Transfer Room	300
	4. Growth Rooms (3 rooms)	250 each
	5. Change Area	100
6.	Semi-Clean Area	
	1. Legwash	100
	2. Laboratory/Media Preparation/Auto Clave	400
	3. Wash Area: (i) Bottle (ii) Plant	200 each
	4. Store (consumables)	250
7.	Service Area	
	1. Office Lobby, Corridor	550
	2. Scientist Room	200
	3. Computer Room	100
	4. Genset Room	150
	5. Toilet	150

Culture media: The medium in which plant tissue grows is made up of various salts (containing all the major and micro elements essential for growth of plants), vitamins, sugars (usually sucrose) and growth regulators at appropriate concentration. Of the various constituents of the medium, the concentration of growth regulators is critical. The plant growth is virtually controlled by the ratio between two groups of growth regulators. Cytokinin group favours shoot growth, whereas auxin group favours root growth. The ratio varies between species and even between varieties within a species.

Sugar is the source of energy for tissues and shoots, while in the laboratory do not normally photosynthesize. Other nutrients perform their usual structural, functional and catalytic role. The agar is added to solidify the medium.

An example of organic culture media is given below:

Organic Compounds	mg/l
Thiamine HCI	0.1
Inositol	100.0
Nicontinic Acid	0.5
Pyridoxine HCI	0.5
Glycine	2.0
Glusose/Sucrose	--
Agar	--
Auxins (IBA/NAA)	--
Cytokinins (BAP/Kinetin)	--

Equipments: Propagation by tissue culture needs a good number of laboratory equipments. The various equipments and their functions are outlined below:

1. *Autoclave (2 nos.):* Sterilisation of all glass apparatus and culture media can be accomplished by means of steam generated in the autoclave.
2. *Analytical/Top Pan balances (2 nos.):* For accurate measurement of various constituents of culture media, these balances are required. Top pan balance is used for measuring larger quantities, while analytical balance is used for measuring smaller quantities.
3. *pH meter (2 nos.):* It is used for measuring and adjusting pH of the culture media or solution.
4. *Laminar Airflow cabinets (2 nos.):* In these cabinets shoots developed on explants are separated from clusters and transferred to fresh medium under sterile condition. Inoculation can also be done here.
5. *Distillation sets (2 nos.):* Water to be used for preparation of culture media should be free from all impurities and salts. This can be accomplished by double distillation of water.
6. *Computer System:* Computerisation of laboratory would be helpful for production planning, time scheduling of subculturing, quality control of plantlets, growth room status, material requirement and market planning.

7. *Air Conditioners and Stabilizers (1t and 1.5t-6 each):* Maintenance of desired temperatures in growth room, inoculation room/culture transfer room by air conditioning.
8. *Microscopes:* Stereo Microscope would enable dissecting out small size meristem from shoot tips by removing the protective covers of leaf primodia. Compound microscope enables detection of bacteria and fungi in culture and plant tissues.
9. *Bottle Washing Unit (1 set)*: Since a large number of bottles or vessels in which plants will be grown are required to be washed repeatedly before use, an automatic bottle washing unit would be helpful.
10. *Media Cooking Unit (1 set):* Culture media, which contains all the essential nutrients, sugar and agar, needs to be cooked before use. A media cooking unit for a large scale commercial unit is, therefore, desirable.
11. *Growth room racks (18 set @ 6 racks/room):* These hold the culture bottles in trays. They are mobile over a set of rails in order to maximise utilisation of space.
12. *Trays (600 @ 4 trays/shelve):* Supporting structure for culture bottles/vessels.
13. *Hatches:* Pass through boxes used as gateway between clean area and semi-clean area for exchanging materials.
14. *Tube lights (600):* Fluorescent tube lights are mounted on the bottoms of the shelves so that culture bottles containing explants/growing tissues receive requisite intensity of lights.
15. *Dissecting kits and Inoculation instruments (5 set):* These are necessary for separation of shoots and preparation of micro cuttings.
16. *Green House:* In tissue-cultural propagation, green house may be required primarily to raise and maintain mother plants so that growth of organs suitable for tissue culture is maximum. It also required to harden the plantlets gradually in natural environment. Green house will enable the

control over light intensity and humidity, which is necessary for hardening of plants. However in case of 100% export-oriented projects, it may be possible to export the plantlets directly from the laboratory without subjecting them to hardening.

17. *Electricity:* No tissue-culture laboratory can operate without electricity. It is required to illuminate tissues and shoots while they are in laboratory, to operate various equipments and facilities like air conditioners, etc. A diesel genset (62.5 KVA) is required to backup in case of irregular electric supply.
18. The laboratory office should have facilities such as Fax Machines, Telephone, Typewriters etc.

Other equipments required for the Tissue-culture Laboratory are:

Sl. No.	Equipments	Quantity
1.	Trolleys	8
2	Generator	1 set
3	Refrigerator	1
4	Air filters	2
5	Oven	1
6	Rotary Shaker	2
7	Bottles	30,000
8	Lab clothes	-
9	Washing Machine	1
10	Incinerator	1
11	Fire fighting equipment	-
12	Miscellaneous Glassware	-
13	Office and Lab. Furniture	-
14	Cupboard and Lab. racks	-
15	Pick up Van	

Water: Water is essential for growing mother plants, hardening of plantlets, washing, distilled water for preparation of culture media and reagents, canteen, toilets, etc. Water supply system should be equipped with:

- Tube wells - 2 nos.
- Overhead Tank (1000 litres)
- Pump House with Pumpset (3 HP)
- Mist system for Green house

Raw materials: Raw materials required for tissue-culture project, apart from explant, are various constituents of culture media. These have already been discussed under paragraphs section 2.1.

Skilled Manpower: Tissue culture is a highly skilled operation. It would, therefore, be essential that laboratory and green house workers are well qualified and experienced in the technology. Their training in well established commercial laboratories would be helpful. All the requirements of tissue-culture projects, mentioned above could be made available from local resources.

2.2 Unit size

The size of tissue-culture project could be expressed in terms of the capacity of production of tissue-cultured plants (TCPs). The projects so far set up in our county with assistance from financial institutions have production capacities, which vary from 1 million to 20 million TCPs per year. The size envisaged in the present model is 1.25 million TCPs per year. It has been estimated that, to produce 1.25 million TCPs, a laboratory of 5000 sq. ft. would be required. A green house facility of 5000 sq. ft. for maintenance of mother plants and hardening tissue cultured plants would be helpful.

2.3 Estimated cost, production and income

The estimated cost for production of 1.25 million plants is indicated below:

No.	Particulars	Cost (Rs. Lakhs)
A	Fixed Cost	
1.	Tissue culture laboratory including green house	19.05
2.	Laboratory Equipments	27.29
3.	Furniture, Fixtures and Office Equipments	6.26
4.	Water Supply System	0.63
5.	Training	1.00
6.	Consultancy/Know-how fees	2.00
7.	Misc.	3.77
B	Recurring Cost	25.00
C	Total Cost	85.00

Estimated production:

Sl. No.	Item	Details
1	Total No. of culture bottles (200 bottles/shelve)	28800
2	No. of bottles to be used for shoot multiplication/cycle	25000
3	No. of shoots to be produced/cycle	125000
4	Total production of shoots per year	1250000
5	Total exportable production (assuming 80%)	1000000

Estimated Income:

Year	Total Production (No.)	Sale (Nos.)	Gross Income (Rs. lakhs)
1	10000	NIL	NIL
2	1000000	750000	37.50
3 onwards	1250000	1,000000	50.00

Financial Assistance: The tissue-culture export-oriented projects are eligible for refinance support by NABARD. Banks may provide loan for the activity provided the scheme is technically feasible and financially viable.

2.4 Tissue culture technology

Tissue culture technology is based on the theory of totipotency i.e. the ability of a single cell to develop into whole organism.

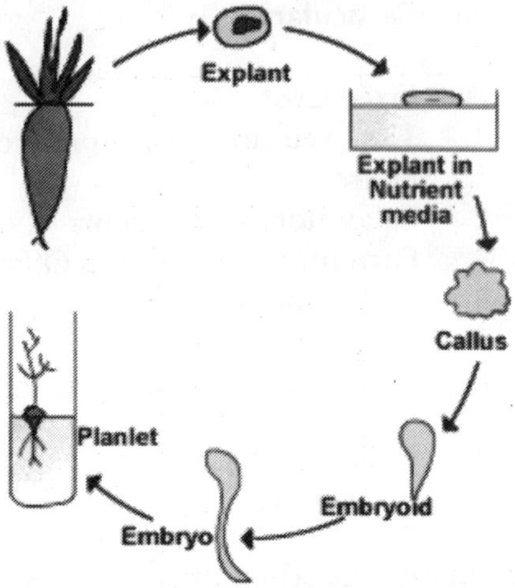

The major components of the technology include choice of explant (excised part of plant), growing of explant on a defined medium in glass vessel (in vitro), elimination and/or prevention of diseases, providing appropriate cultural environment and transfer of plantlets from glass vessel to natural environment (hardening). All these constitute protocol for tissue culture. It varies from species to species and variety to variety within the same species.

2.4.1 Stages of tissue culture

The propagation of plant by tissue culture are dividend into five stages. A general account of these stages is outlined below:

Choice of explants: Explants could be shoot tips (meristem), nodal buds, sections from internodes, leaves, roots, centres of bulbs, corms or rhizomes, or other organs. The choice depends on the species to be multiplied and the method of shoot multiplication to be followed. Activity like growing (shoot tips), juvenile (seedlings) or rejuvenated (suckers) tissues are preferred.

The commercial tissue culture laboratories commonly use tips of apical or lateral shoots, which contain meristems. Meristems are made up of cells dividing actively in an organised manner. They are about 0.1 mm in diameter and 0.25 - 0.30 mm in length. However, explants should be chosen from typical, healthy, disease free, well-tested mother plants cultivated under conditions which reduce contamination and promote growth of tissues to be cultured. If necessary explants may be subjected to virus testing and elimination. The selection of mother plants is very important for commercial success of tissue culture propagation.

The quantity of explant required for propagation by tissue culture is very small. For example, 2 mm thick petiole sections from African violet (a flowering herb) could yield 20,000 plantlets per petiole (basal portion of leaf).

Establishment of germfree (aseptic/sterile) culture: Excised part of plant is surfaced sterilised and transferred to sterile nutrient medium contained in glass vessel. On an average, about 50 cc nutrient medium may be added per glass vessel. The cultures are maintained in growth rooms. If there is no infection and tissue isolated from mother plants survive in the artifical environment, initiation of new growth will take place after a week or so. Thus, germ-free culture is established.

Production of shoots/propagules: Once growth is initiated by induction of meristematic centres, buds develop into shoots by multiplication of cells. There are three types of multiplication systems for production of shoots.

i) *Multiplication by axillary shoots:* In this case shoots are produced from excised shoot tips or nodes. Hormones (cytokinins) are used to induce multiple branching. This is the most common method followed in commercial units. However, the rate of multiplication is low. Still it is preferred, because axillary shoots are likely to be genetically stable and the chances of production of types unlike mothers are less.

ii) *Multiplication by adventitious shoots:* Explants such as sections of leaves, internodes or roots can produce directly adventitious shoots or other organs. This system has higher multiplication rate, but lesser genetic stability than axillary system.

iii) *Multiplication by somatic embryos (embryoids):* Embryos are usually formed by the union of male and female reproductive cells (zygotic embryo) which ultimately can develop into a young plant. Embryo - like structures can also be produced from somatic cells. Somatic embryos are independent bipolar structures and are not attached to the tissues of origin. They also can develop to form young plants like zygotic embryos. Somatic embryos may be produced directly from explants such as sections of leaves, internodes or roots on solid culture medium.

In certain species, the formation of young plants, or formation of somatic embryos occurs directly on excised plant parts. The most common form of regeneration of plants occurs indirectly from callus. Callus is a mass of undifferentiated dividing cells often formed in tissues cultured in vitro. Callus may give rise either to adventitious shoots, which develop into plantlets, or somatic embryos, which develop into seedlings. Callus is formed even naturally in response to wound.

The formation of callus can be induced by selecting proper tissue and culture medium. This system has the highest multiplication rate and produce complete tiny plants. One gram of explants can produce one lakh somatic embryos. Dormancy can be induced in them or they can be transformed into synthetic seeds. However, callus is genetically unstable or plants arising from it may be unlike mother plants. Such plants are known as off-types. They occur more frequently in callus culture and adventitious shoot culture as compared to axillary shoot culture. Off-types are undesirable in commercial propagation. Regeneration of shoots or intact plants by any one of the multiplication systems described above is influenced by many factors, such as composition of medium (specially concentration of growth regulators), type of tissue, genotpye, ploidy level, etc.

Normally, multiplication cycle i.e., the period from incubation of plant parts on medium to formation of shoots varies from 3 to 6 weeks. However, the process is recycled many times by sub-culturing in order to obtain required multiplication rates. After completion of a cycle, shoots are cut separately and transferred to fresh medium. Cutting is done manually by using dissecting tools in laminar flow cabinets, where the air is clean to prevent any contamination. Once the shoots are placed on fresh medium, they are transferred back to the growth rooms. Thus, it may be possible to multiply the shoots 3 to 10 times per cycle of 3 to 6 weeks duration.

Preparation of micro-cuttings for establishment in the natural environment: Young axillary or adventitious shoots are finally separated form clusters (micro-cutting) for initiation and development of roots. After separation, they are transferred individually to a medium containing rooting hormone (auxin) and continued to be maintained in the growth rooms until the roots are formed. It may also be possible to transfer the micro cuttings directly to soil or compost in humid green house for root formation. Somatic embryo's may directly develop into seedlings.

Establishment in the natural environment: The most critical stage of the propagation by tissue culture is the establishment of the plantlets into the soil. The steps involved are as under:

- Washing of media from plantlets
- Transfer of plantlets to compost/soil in high humid green house
- Gradual decrease in humidity from 100% to normal over 3-4 weeks
- Gradual increase in light intensity.

Plantlets during their growth in laboratory do not photo synthesise and their control of water balance is very weak. They use sugar contained in medium as source of energy. They exist like bacteria (heterotrophy). They need to be converted to more plant like existence (autotrophy) i.e., they should be in a position to utilise carbon dioxide from the air and solar energy for their food requirement. This acclimatization on the harsh real environment outside artificial laboratory takes place gradually.

2.4.2 Culture environment

Environment conditions in the growth room which influence cell multiplication are light, day length and temperature. In tissue culture, light is required for synthesis of green pigment (chlorophyll) and development of organs. The range of light intensities appropriate for culture room varies from 1000 to 5000 lux. Requirement of day length would be in the range of 16-18 hours. Temperature requirement varies from 20-30°C depending on species of plants. Tropical plants may require higher temperature.

2.4.3 Prevention of contamination

Prevention of contamination in tissue culture is extremely important for commercial success of the unit. The entire production can go waste if the culture is contaminated. Sugar rich culture medium, excised plant tissue and culture environment are all conducive to the growth of pathogens. Therefore, it is essential that all operations should be conducted in sterile or aseptic conditions. Various stages involved in prevention of contamination are outlined below:

- Mother plants should be grown under conditions which do not promote diseases. Explants should be free of disease. Meristem is usually free from disease.
- Surface sterilisation of explants in solutions of sodium or calcium hypochlorite is necessary. Heat or treatment with certain chemicals may eradicate latent viruses.
- All equipments and culture media are sterilised by autoclaving at 15 lb/sq. inch pressure at 120° C for 15 minutes.
- The laboratory should be cleaned with disinfectants. Workers should wash their hands and feet with disinfectants before entering the laboratory. They should put on sterilised clothes.
- Double distilled water should be used for washing explant and preparation of culture medium.
- UV lamps assist in sterilisation of laminar flow cabinets, hatches and instruments.

Air handling units are employed for growth rooms, and culture transfer rooms in order to avoid cross contamination between different areas of operation inside the clean area. The sterile condition is obtained in laminar air flow cabinets as they are provided with special type of international standard HEPA filters. These filters remove all the dust particles of above 0.3 micron present in the air.

Chapter 9

Organic Input Evaluation

Regulation of plant protection products (PPP) and fertilizers and soil conditioners (F&SC) has great implications on farming practices and on the crop-specific economy, and hereby on the income and the competitiveness of farmers. The strict regulation of PPP and F&SC in organic agriculture is also of crucial importance for the trust of the consumers in the safety and quality of certified organic crops.

The objective of the Organic Inputs Evaluation is to evaluate farm inputs and confirms their compliance with the Indian National Organic Standards and other country's standards like US-NOP, Council Regulation EEC 2092/91, JAS, etc. Presently input approval scheme is applicable to fertilizers and soil conditioners and to inputs related to plant protection (pesticides, repellents etc).

1. Evaluation of Organic Inputs

When an input is to be evaluated it must first be investigated by certification programmes to see whether it fulfils the required criteria. An input must fulfil all 6 requirements before it can be accepted as suitable for use in organic agriculture.

i) **Necessity:** The necessity of each input must be established. This will be investigated in the context in which the product will be used. Arguments to prove the necessity of an input may be drawn from such criteria as yield, product quality, environmental safety, ecological protection, landscape, human and animal welfare. The use of an input may be restricted to *specific crops* (especially perennial crops), *specific regions* and *specific conditions* under which the input may be used.

ii) Natures and method of production: The origin of the input should usually be organic (vegetative, animal, microbial). When there is any choice, renewable inputs are preferred. The next choice is inputs of mineral origin and the third choice is inputs which are chemically identical to natural products. There may be ecological, technical or economic arguments to take into consideration in the allowance of chemically identical inputs.

Method of production: The ingredients of the inputs may undergo the processes like mechanical, physical, enzymatic, action of microorganisms and chemical (as an exception and restricted)

Collection: The collection of the raw materials comprising the input must not affect the stability of the natural habitat nor affect the maintenance of any species within the collection area.

iii) Environment: The input must not be harmful or have a lasting negative impact on the environment. Nor should the input give rise to unacceptable pollution of surface or ground water, air or soil. All stages during processing, use and breakdown must be evaluated. The following characteristics of the input must be taken into account:

Degradability: All inputs must be degradable to their mineral form. Inputs with a high acute toxicity to non-target organisms should have a maximum half-life of five days. Natural substances used as inputs, which are not considered toxic, do not need to be degradable within a limited time.

Acute toxicity to non-target organisms: When inputs have a relatively high acute toxicity for non-target organisms, a restriction for their use is needed. Measures have to be taken to guarantee the survival of these non-target organisms. Maximum amounts allowed for application may be set. When it is not possible to take adequate measures, the use of the input must not be allowed.

Long-term chronic toxicity: Inputs which accumulate in organisms or systems of organisms and inputs which have, or are suspected of having, mutagenic or carcinogenic properties must not be used. If there are any risks, sufficient measures have to be taken to reduce any risk to an acceptable level and to prevent long lasting negative environmental effects.

Chemically synthesised products and heavy metals: Inputs should not contain harmful amounts of man made chemicals (xenobiotic products). Chemically synthesised products may be accepted only if identical to the natural product.

Mineral inputs should contain as few heavy metals as possible. Due to the lack of any alternative, and long-standing, traditional use in organic agriculture, copper and copper salts are an exception for the time being. The use of copper in any form in organic agriculture must be seen, however, as temporary and use must be restricted with regard to environmental impact.

iv) Human health and quality: Inputs must not be harmful to human health. All stages during processing, use and degradation must be taken into account. Measures must be taken to reduce any risks and standards set for inputs used in organic production.

Inputs must not have negative effects on the quality of the product - e.g. taste, appearance and quality.

v) Ethical aspects - animal welfare: Inputs must not have a negative influence on the natural behaviour or physical functioning of animals kept at the farm.

vi) Socio-economic aspects: Inputs should not meet resistance or opposition of consumers of organic products. An input might be considered by consumers to be unsafe to the environment or human health, although this has not been scientifically proven. Inputs should not interfere with a general feeling or opinion about what is natural or organic - e.g. genetic engineering.

The approval is based on an evaluation of the product documentation and on an inspection of the processing unit. Applicants need to submit a complete list of ingredients of the product and a list of processes involved in the manufacturing. They need to grant access to Organic Inspectors to the processing unit and the book keeping related to the product.

Inputs should be evaluated regularly and weighed against alternatives. This process of regular evaluation should result in organic production becoming ever more friendly to humans, animals, environment and the ecosystem.

2. How to apply for input approval?

Before applying for input approval, organic operator should self-assess whether all ingredients of your products are permitted according to the organic standards (Appendix 1 & 2) in order to avoid unnecessary efforts and costs. To apply for input approval, application form may be procured from the Organic Certifying Agency. Complete the input approval application form and send it to the Certifying Agency Office. One application can include several products at the same time. Based on the application form evaluation is done.

List of Certifying Agency working in India for certification of organic farms and organic input evaluation are given below:

1. *Indian Organic Certification Agency (INDOCERT), C/o* Executive Director, Thottumugham P.O. Aluva-683105, Ernakulam Kerala State, India
2. *SGS India Pvt. Ltd.* C/o Business Manager, M/s SGS India Pvt. Ltd., 250 Udyog Vihar, Phase - IV, Gurgaon - 122015
3. *Skal International (Netherlands),* International Inspector, Skal Inspection and Certification Agency, No. 191, 1st Main Road, Mahalaxmi Layout, Bangalore - 560086
4. *APOF Organic Certification Agency,* 1st Floor, 5th Main, 9th Cross, Jayamahal Extension, Banglore-46
5. *OneCert Asia Agri Certification Pvt. Ltd,* Agrasen Farm, Vatika Road, Off. Tonk Road (P.O.Vatika), Jaipur - 303 905, Rajasthan, India.
6. *Bioinspecta,* C/o INDOCERT, Thottumugham P.O. Aluva-683105, Ernakulam, Kerala State, India
7. *Ecocert International,* Country Representative/Managing Director, Ecocert SA Branch Office, 54 A, Kanchan Nagar, Nakshetrawadi Aurangabad - 431002, Maharashtra State

8. *IMO Control Private Limited,* Director, 26, 17th Main, HAL 2nd 'A', Stage, Bangalore-560 008.

3. Information required for Input Approval

1. Brief description about Input manufacturing company:
2. No. of production units for input manufacturing:
3. Location of these production units:
4. List all products for approval:
   - Product Name
   - Type of input
     - ❑ Fertilizer / soil conditioner
     - ❑ Pesticide / Repellent
     - ❑ Other:
5. If your product has been registered under government regulation, provide registration number:

   Product name

   Registration no.
6. Any subcontracted service. If yes, please list all companies (along with address) that further process/package of products
7. List all storage facilities for raw material and finished products:
8. Location name/address of storage
9. Type of storage
10. Materials stored
11. Responsible Person
12. Methods of cleaning of the storage facilities:
13. Methods used for cleaning and sterilizing the processing unit:
14. Name of the quality assurance system followed in unit:
15. Any pest problems being faced in the processing unit or storage room? If yes, list the kind of pest and the remedial measures taken to control them.

16 List the kind of packaging material used:
   - Product
   - Packaging material
17. Continuous book keeping availability
18. In which way is the intake of goods documented?
19. In which way the goods issued are documented (outward movement)
20. Is it possible to trace back every single lot processed

Document Required for verification

- Complete list of ingredients (including sources)
- Complete list of processes
- Government registration documents
- Chemical analysis report
- Contract with subcontractor
- Site maps
- Flow chart for processing
- Receipt / invoice of all ingredient
- Receipt of all sold product
- Sample of all packaging materials
- Import certificates for the imported products

Frequently Asked Questions

What is organic farming?

Cultivating the land for raising field crop using biological sources of plant nutrients without involving any chemical either as fertilizer or Insecticides to avoid its possible ill effects on soil, ground water, crop and ecology.

How do organic farmers fertilize crops?

In organic farms to build healthy soils farmer has to nourish the living component of the soil, the microbial inhabitants that release, transform, and transfer nutrients. Soil organic matter contributes to good soil structure and water-holding capacity. Organic farmers feed soil biota and build soil structure and water-holding capacity. Organic farmers feed soil biota and build soil organic matter with compost, vermicompost, cover crops, and biologically based soil amendments. These produce healthy plants that are better able to resist disease and insect predation.

How do organic farmer control pests and diseases?

Strategy in controlling pests and diseases in organic farms is prevention through good plant nutrition and management. Organic farmers have to rely on a diverse population of soil organisms, beneficial insects, and birds to keep pests in check. When pest populations get out of balance, growers has to implement a variety of strategies such as the use of insect predators, mating disruption, traps and barriers. As a last resort, certain botanical or other non-synthetic pesticides may be applied. Under the National Organic Rule, growers are required to use sanitation and cultural practices first before they can resort to applying a material to control a weed, pest or disease problem. Use of these materials in organic production is regulated, strictly monitored, and documented.

What are organic manures?

Organic manures are natural products used by farmers to provide food (plant nutrients) for the crop plants. There are a number of organic manures like farmyard manure, green manures, compost prepared from crop residues and other farm wastes, vermicompost, oil cakes, and biological wastes - animal bones, slaughter house refuse.

How are organic manures beneficial in the cultivation of crops?

Organic manures increase the organic matter in the soil. Organic matter in turn releases the plant food in available form for the use of crops. However, organic manures should not be seen only as carriers of plant food. These manures also enable a soil to hold more water and also help to improve the drainage in clay soils. They provide organic acids that help to dissolve soil nutrients and make them available for the plants.

How are organic manures differing from fertilizers?

Organic manures have low nutrient content and therefore need to be applied in larger quantities. For example, to get 25 kg of NPK, one will need 600 to 2000 kg of organic manure. The nutrient content of organic manures is highly variable from place to place and method of preparation.

How much of plant nutrients are provided by organic manures?

Just as different fertilizers contain different amounts of plant nutrients, organic manures are also not alike. Average quality of farmyard manure provides 12 kg nutrients per ton and compost provides 40 kg per ton. Most of the legume green manures provide 20 kg of nitrogen per ton. Each ton of sorghum/rice/maize straw can be expected to add 26 kg of nutrients.

How a good compost is prepared?

Compost making is the process of decomposing organic wastes in a pit. Site for compost making is selected should be at a high level and water should not pond during monsoon season. The composting materials include cattle dung, plant residues, waste fodder, dried plants stalks and leaves, etc. These materials should be placed in layers inside the pit. Close the pit with urine earth, waste fodder and then heap the soil till it gets convex shape (about 1 to 1.5′ above the ground) so that the rainwater rolls away. After six months compost is ready to apply to the fields.

What is vermicomposting?

Vermicomposting is a type of compost making in which earthworms are used to convert organic wastes into valuable material to supply nutrients for crops. Organic wastes are buried in the worm pit and worms eat the food waste, bedding and bacteria. They turn it all into humus nutrient-rich food for growing healthy plants. The brown, rich soil they produce is called worm castings.

When is the vermicompost ready?

The compost is ready when the material is moderately loose and crumbly and the colour of the compost is dark brown. It will be black, granular, lightweight and humus-rich. Any bad odor is a sign that fermentation has not reached its final goal and that the bacterial processes are still going on.

How do I separate the worm from vermicompost?

To facilitate separating the worms from the ready vermicompost, stop watering two to three days before emptying the beds. This will force about 80 per cent of the worms to the bottom of the bed. The rest of the worms can be removed by hand, and are ready to be transferred into the next round of compost making.

How do I harvest the vermicompost?

Harvesting the vermicompost can be done several ways, but the most popular two ways are- dumping/hand sorting and side harvesting. Dumping/hand sorting is done by first preparing new bedding. Then, the old bin material is turned onto a large sheet of plastic and a bright light focused on the top of this pile. Since the worms are photosensitive, they burrow away from the light. The top of the pile can be scraped away. The worms uncovered will then burrow away from the light so the top can again be scraped away. Repeat this process until most of the vermicompost has been harvested.

Side harvesting is accomplished by only feeding the worms on one side of the bin for a few weeks. The worms will all migrate to that side. Then harvest the vermicompost from the unoccupied side of the bin. Put new bedding in the harvested side and feed the worms only on that side for a few weeks. The worms will then migrate to the new side so the remaining vermicompost in the old side can be retrieved.

How often do I harvest the castings?

The vermicomposting process takes two to four months. After about six weeks of the worms eating scraps in your bin/pit, you will begin to see worm castings. They will have an appearance of dark granules. After eight weeks, the vermicompost should be ready for harvesting.

What problems might occur in vermicomposting and how do I solve them?

Some vermicomposting problems that can occur are odor, flies or other pests, worms escaping from bin/pit, mold forming on bedding, bedding drying out, and water collecting in the bottom of the bin/pit. These problems can be easily eliminated. There will be no odor or pests, if food is properly buried. Worms will not escape, if screening is on bottom of bin to cover ventilation holes. Mold will not form, bedding will not dry out and water will not collect at the bottom of the bin, if the bin/pit is kept appropriately moist.

What should I feed my worms?

Just about any non-dairy and non-meat kitchen scraps will be perfect for your worm bin/pit. Dairy and meat products can attract pests and make your compost greasy. Egg shells are good to add to your bin/pit since they provide calcium for the worms. You will see your worms curled up inside the egg shells. Coffee grounds, coffee filters and tea bags are fine too. Dried leaves can also be fed to the worms. Earthworms shy away from garlic, onions, and any citrus fruit or citrus peel.

What is biofertilizer?

Biofertilizers are ready to use live formulates of such beneficial microorganisms like bacteria, fungi and algae which on application to seed, root or soil mobilize the availability of nutrients by their biological activity in particular, and help build up the micro-flora and in turn the soil health in general. Biofertilizers are capable of fixing atmospheric nitrogen or solubilising insoluble phosphate in soil and making them available to the crop plants.

Why biofertilizers are environmental friendly?

The biofertilizers are not at all harmful to soil, predators, animals and human beings. Moreover they are pollution free and renewable. Hence they are called environmental friendly.

Why should we use biofertilizers?

With the introduction of green revolution technologies the modern agriculture is getting more and more dependent upon the steady supply of synthetic inputs (mainly fertilizers), which are products of fossil fuel (coal+petroleum). Adverse effects are being noticed due to the excessive and imbalanced use of these synthetic inputs. This situation has lead to identifying harmless inputs like biofertilizers. Use of such natural products like biofertilizers in crop cultivation will help in safeguarding the soil health and also the quality of crop products.

What are the advantages of biofertilizer?

Fixes atmospheric nitrogen or solubilises insoluble phosphates in the soil and provides ever increasing biological nitrogen to the plants. It enhances germination and plant growth due to release of vitamins, auxins and harmones, increased yield by 20-30%. It also controls and suppresses soil borne diseases to some extent (Antagonise) and helps in survival of beneficial microorganisms in the soil (Proliferate).

Whether biofertilizer can supply all the three major plant nutrients?

No. At present, biofertilizers are made available for nitrogen and phosphorus only. No biofertilizer is so far available for potassium.

Can one biofertilizer supply two major plant nutrients?

No. One biofertilizer can supply/made available mainly one major nutrient.

What nutrient is supplied by the algal group?

The algal group supplies only nitrogen.

What nutrient is supplied by the fungal group?

The fungal group solubilises insoluble forms of phosphate present in the soil and make it available to the crop plants.

What nutrient is supplied by the bacterial group?

The bacterial organisms present in the biofertilizer either fix atmospheric nitrogen or solubilises insoluble forms of soil phosphate.

What is the most important source of N?

It is available in the atmosphere. The atmospheric air contains about 79% nitrogen in gaseous form. One hectare area column of atmospheric air contains approx. 80,000 t of nitrogen. This form of nitrogen (N_2) from air cannot be utilised by plants as such.

What is Rhizobium?

Nitrogen is available to the leguminous plants mainly through biological nitrogen fixation by the root nodule bacteria called Rhizobium. These bacteria are symbiotic in nature and host specific. Higher yields in legumes can be obtained by exploiting this system.

What is symbiotic association?

Certain bacteria like Rhizobium live inside the root nodules of leguminous plants. These nodules are bacterial houses. While living inside the root nodules, the bacteria get shelter and food material from the plant and fix atmospheric nitrogen which is used by the plants. The plants and bacteria both are mutually benefited and hence it is called symbiotic association.

Are the usages of bacterial strains it for all the leguminous crops?

No. Cowpea Rhizobium benefits certain other legumes also whereas other Rhizobium strains have specific hosts. It is necessary to apply only the specific strain, which is recommended for that crop.

What is Acetobacter?

Acetobacter is symbiotic bacteria capable of fixing atmospheric nitrogen by living within the sugar plant. The organism is found in all parts of plant body. The *Acetobacter* is suitable for sugarcane cultivation.

What is an Associative Symbiotic Bacteria?

This bacterial group live partly within the root and partly outside. There is a fair degree of symbiosis between the host and the bacteria. Hence, they are called as Associative Symbiotic Bacteria. Azospirillum is an important bacterium in this group, recommended for millets, grass, wheat, maize, sorghum, rice etc.

What is a non-symbiotic bacterium?

Certain bacteria live independent of root system of plant capable of fixing nitrogen or solubilising soil phosphate without any symbiotic association and hence they are called non-symbiotic bacteria or free-living symbionts.

What is Azotobacter?

It is non-symbiotic nitrogen fixing bacteria, aerobic in nature, recommended for non-leguminous crops like paddy, millets, cotton, tomato, cabbage and other monocotyledonous crops. Azotobacter also produces (VAH) growth promoting substances like IAA, Gibberellic acid, Cytokinin, Vitamins and certain chelating agent and polysaccharides as reducing and binding agents. Azotobacter performs well, if the soil organic matter content is high.

What biofertilizers are recommended for crops?

Rhizobium + Phosphotika at 200 gm each per 10 kg of seed as seed treatment are recommended for pulses such as pigeonpea, green gram, black gram, cowpea, groundnut and soybean.

Azotobacter + Phosphotika at 200 gm each per 10 kg of seed as seed treatment are useful for wheat, sorghum, maize, cotton, mustard etc.

For transplanted rice, the recommendation is to dip the roots of seedlings for 8 to 10 hours in a solution of Azospirillum + Phosphotika at 5 kg each per ha.

What is Azolla?

Azolla is an aquatic floating fern, found in temperate climate suitable for paddy cultivation. The fern appears as a green mat over water, which becomes reddish due to excess anthocyanin pigmentation. The BGA cyanobacteria (*Anabaena azollae*) present as symbiont with this fern in the lower cavities actually fixes atmospheric nitrogen. The rate of nitrogen fixed is around 20-25 kg/ha.

What is the dose of Azolla required for one-acre paddy crop?

Azolla application can be done in two ways: one as green manure, where *Azolla* is grown alone (two to three weeks) in flooded fields, water drained and *Azolla* fern is incorporated (10 mt material) in the field before planting paddy. Second method 4-5 q of fresh *Azolla* is applied in standing water one week after planting of paddy. When a thick mat of *Azolla* is formed, water is drained and *Azolla* is incorporated into the soil.

What is blue green algae?

The blue green algae are also called as cyanobacteria. This Chlorophyll containing algal organism fixes atmospheric nitrogen. Application of BGA (10 kg/ha) is recommended for flooded paddy as it can survive and multiply easily in standing water.

What are the important phosphate solubilising organisms?

1. *Pseudomonas striata*
2. *Bacillus Polymyxa/megaterium*
3. *Aspergillus awamori*
4. *Penicillium digitatum*

How the phosphate solubiliser is functioning in the soil?

The phosphate solubiliser produces organic acids like tartaric, fumeric, malic, succinic and acetic acid etc. which solubilise insoluble forms of phosphate present in the soil to available form.

Whether the phosphate solubilisers are crop specific?

No. They can be applied to and recommended for all crops.

How could one get good response to biofertilizer application?

- Biofertilizer product must contain good effective strain in appropriate population and should be free from contaminating microorganisms.
- Select right combination of biofertilizers and use before expiry date.
- Use suggested method of application and apply at appropriate time as per the information provided on the label.
- For seed treatment adequate adhesive should be used for better results.
- For problematic soils use corrective methods like lime or gypsum pelleting of seeds or correction of soil pH by use of lime.
- Ensure the supply of phosphorus and other nutrients.

What would be probable reasons for not getting response from the application of biofertilizers?

1. On account of quality of product
    - Use of ineffective strain.
    - Insufficient population of microorganisms.
    - High level of contaminants.
2. On account of inadequate storage facilities
    - Exposed to high temperature.
    - Stored in hostile conditions.
3. On account of usage
    - Not used by recommended method in appropriate doses.
    - Poor quality adhesive.
    - Used with strong doses of plant protection chemicals.
4. On account of soil and environment
    - High soil temperature or low soil moisture.

- Acidity or alkalinity in soil.
- Poor availability of phosphorous and molybdenum.
- Presence of high native population or presence of bacteriophages.

What is VAM?

The VAM is Vesicular Arbuscular Mycorrhizae, called fungi - which possess special structures known as vesicles and arbuscules - later helps in the transfer of nutrients from soil to root system. These are intercellular, obligate endosymbionts - which have not yet obtained in pure culture. They often help increased uptake of nutrients and water. These fungi (VAM) are found very suitable for groundnut, soybeans, millets, coffee, citrus, pepper, cloves nutmeg etc.

Does VAM act as phosphate solubiliser?

Yes. Mycorrhizae help in mobilize insoluble soil phosphates. They further help increasing nutrient uptake (phosphorus as well as zinc).

What is the method of biofertilizer applications?

Seed treatment: 200 g of nitrogenous biofertilizer and 200 g of Phosphotika are suspended in 500 ml of water and mixed thoroughly. Ten kg seeds are treated with this paste and dried in shade. The treated seeds have to be sown as soon as possible.

Seedling root dip: For rice crop, a bed is made in the field and filled with water. Recommended biofertilizers are mixed in this water and the roots of seedlings are dipped for 8-10 hrs.

Soil treatment: 5kg each of the recommended biofertilizers are mixed in 100 kg of compost and kept overnight. This mixture is incorporated in the soil at the time of sowing or planting. 10 kg Blue Green Algae should be applied for one hectare of wetland paddy. Similarly, 10 qt Azolla is recommended for wetland paddy.

What is seed treatment and how it should be done?

Coating the seed with bio-inoculants is generally known as seed-treatment or bacterisation. It ensures quick germination, fast growth of crop plants, increased yield and better product. It can be done either by dry mix/wet coat or palletizing process.

What is decomposing biofertilizers?

Decomposing biofertilizers are the microbial preparations used to enhance (fast) decomposition of the organic materials both cellulolytic as well as lignolytic and to reduce the bulk size of the finished material.

What precautions one should take for using biofertilizers?

- Biofertilizer packets need to be stored in cool and dry place away from direct sunlight and heat.
- Right combinations of biofertilizers have to be used.
- As Rhizobium is crop specific, one should use for the specifield crop only.
- Other chemicals should not be mixed with the biofertilizers.
- While purchasing one should ensure that each packet is provided with necessary information like name of the product, name of the crop for which intended, name and address of the manufacturer, date of manufacture, date of expiry, batch number and instructions for use.
- The packet has to be used before its expiry, only for the specified crop and by the recommended method of application.
- Biofertilizers are live product and require care in the storage
- Both nitrogenous and phosphatic biofertilizers are to be used to get the best results.
- It is important to use biofertilizers along with organic manures.

Biofertilizers are not replacement of fertilizers but can supplement plant nutrient requirements.

What are Pesticides?

Pesticide is a broad term, covering a range of products that are used to control pests. Slug pellets, ant powder, weed killers, rat and mouse baits are all pesticides. Other pesticides you may have heard of include: insect killers (insecticides), mould and fungi killers (fungicides) , weed-killers (herbicides), slug pellets (molluscicides), plant growth regulators, bird and animal repellents and, rat and mouse killers (rodenticides).

What Legislation deals with Pesticides?

As the bio-control agents are living organisms, it is very important to have effective regulatory measures. Directorate of Plant Protection Quarantine and Storage, Department of Agriculture and Cooperation, Ministry of Agriculture, Government of India has issued guidelines/data requirements for registration of bio-pesticides in the country. As per this, all the units have to meet the Indian standards and technical specifications to be eligible for registration under the Insecticides Act, 1968. This Act regulates the import, manufacture, sale, transport, distribution and use of insecticides with a view to prevent risks to human beings or animals. The provisions under the Act and the Rules are implemented by the Central Insecticides Board and Registration Committee under Ministry of Agriculture at the Central and licencing authorities in the States.

Do Organic, Natural or Bio-pesticides have to be Registered?

At present, *Bacillus thuringensis*, neem based formulations, microbial pesticides like fungi, NPV etc., are included in the schedule of Insecticides Act, 1968. This ensures the quality of bio-pesticides at farmers level. The standard parameters, protocols for data generation, guidelines for registration are prepared and circulated to prospective entrepreneurs by Ministry of Agriculture, Government of India. Any person desiring to manufacture any pesticides is required to register and also obtain licence from the Registration Committee and licencing authority respectively. The Insecticides Rules, 1971 notified under this Act, prescribe certain provisions regarding use of protective clothing, equipment and other facilities for workers during manufacture of insecticides or pesticides. Now as such, any person dealing with biopesticides without registration is illegal.

Guideline for registration of bio-pesticides in India is available at website: http://cibrc.nic.in (Central Insecticide Board and Registration Committee).

What is Tissue culture technology?

Tissue culture means ability of a single cell to develop into whole organism. The major components of the technology include choice of explant (excised part of plant), growing of explant on a defined medium in glass vessel (in vitro), elimination and or prevention of diseases, providing appropriate cultural environment and transfer of plantlets from glass vessel to natural environment (hardening). All these constitute protocol for tissue culture. It varies from species to species and variety to variety within the same species.

What are the advantages of Tissue culture?

A large number of true to the type plants could be propagated within a short time and space and that too throughout the year. Tissue-culture could be a useful way of eliminating disease which can accrue in stock plants. Tissue cultured plants may have increased branching and flowering, greater vigour and higher yield, mainly due to the possibility of elimination of diseases. The method may succeed to propagate plants where seeds or vegetative propagation is not possible or difficult or undesirable. The method saves space and energy. The flexibility of nurseries can be improved. As the capital investment on mother plants is reduced to almost zero, it may be easier to adapt to changing conditions. Additionally, a better programming of the production is possible, because of the greater plant uniformity and the availability in the mass at any time. Tissue culture can be utilised for breeding new varieties, preservation of germplasm and in vitro synthesis of metabolites.

What is the commercial prospect of Tissue culture in India?

Propagation by tissue culture offers good commercial prospect in ornamental plants, vegetables and also fruit plants, where value of the products is high. The technique has reportedly been successful in more than 100 species of plants. It has been estimated that more than 350 million plants are being produced annually through tissue culture. It is well known that one of the major constraints of horticultural development in our country is inadequate availability of quality planting materials. Hence, the demands of planting materials for major fruit crops are very high.

In India there are more than ten commercial organisations, which have developed technical competence for tissue culture with or without foreign tie-ups. The present installed capacity is about 50 million and the export is of the order of Rs. 5 crores. The working group appointed by the Ministry of Commerce, Government of India has proposed an export target of Rs. 30 crores i.e. about 60 million tissue culture plants over a five-year period as against the present production of 8-10 million. Tissue culture method of propagation is highly labour intensive, 55-60% of the cost is on account of labour. India's potential for export of tissue culture plants is rated very high because of abundant and cheap labour.

References

1. Biofertilizer - Technology, Marketing and usage, a source book-cum-glossary (1995) by Dr. M R Motsara *et al*- Fertilizer Development and Consultation Organization Publishers, New Delhi 110048 (India)
2. Biofertilizers in Agriculture and Forestry by N S Subba Rao, Oxford & IBH, New Delhi (India).
3. Biological Nitrogen Fixation (1993) by N.S. Subba Rao, G.S. Venkataraman and S. Kannaiyan, ICAR, New Delhi (India).
4. Composting and Vermicomposting (2001) by S.K. Agarwal and L.M. Saxena, Society for Environment and Development, New Delhi (India)
5. Conversion to Organic Agriculture (2007) by A.K. Singh, International Book Distributing Co., Lucknow, India (2006)
6. Earthworms: Nature's Gift for Utilization of Organic Wastes (1998) by Kale, R. D. In: Edwards, C.A. (Ed.) *Earthworm Ecology*, St. Lucie Press, Washington D.C.
7. Earthworm, vermiculture and vermicomposting (1996) by R.K. Bhatnagar and R.K. Palta, Kalyani Publishers, Ludhiana (India).
8. EU Regulation (EEC) No. 2092/91 and 1804/1999
9. http://www.indocert.org (Indian Organic Certification Agency)
10. ISI Standard IS: 8268-1986 (Specifications for Rhizobium)
11. ISI standard IS: 9138-1979 (Specifications for Azotobacter)
12. NABARD website
13. National Center of Organic Farming (NCOF) website
14. National Programme on Organic Production (NPOP), May 2005, APEDA, Ministry of Commerce, Govt. of India.
15. Utilization of earthworms in recycling of household refuse-a case study (1993). Kale, R.D. and Sunita, N.S. In: CORT (Ed.) *Biogass slurry utilization*, New Delhi
16. Vermiculture and Organic Farming (2004) by T.V. Sathe, Daya Publishing House, India

Appendices

APPENDIX 1

PERMITTED/ RESTRICTED PRODUCTS UNDER CERTIFIED ORGANIC FARMING (source: www.apeda.com)

Products for Use in Fertilising and Soil Conditioning

MATTER PRODUCED ON AN ORGANIC FARM UNIT	
Farmyard & poultry manure, slurry, urine	Permitted
Crop residues and green manure	Permitted
Straw and other mulches	Permitted
MATTER PRODUCED OUTSIDE THE ORGANIC FARM UNIT	
Blood meal, meat meal, bone meal and feather meal without Preservatives	Restricted
Compost made from any carbon based residues (animal excrement including poultry)	Restricted
Farmyard manure, slurry, urine (Preferably after control fermentation and /or appropriate dilution) "Factory" farming sources not permitted.)	Restricted
Fish and fish products without preservatives	Restricted
Guano	Restricted
Human excrement	Not Allowed
By-products from the food and textile industries of biodegradable material of microbial, plant or animal origin without any synthetic additives	Restricted
Peat without synthetic additives (prohibited for soil conditioning)	--
Sawdust, wood shavings, wood provided it comes from untreated wood	--
Seaweed and seaweed products obtained by physical processes, extraction with water or aqueous acid and/or alkaline solution	Restricted

Sewage sludge and urban composts from separated sources which are monitored for contamination	Restricted
Straw	Restricted
Vermicasts	Restricted
Animal charcoal	Restricted
Compost and spent mushroom and vermiculate substances	Restricted
Compost from organic household reference	Restricted
Compost from plant residues	
By products from oil palm, coconut and cocoa (including empty fruit bunch, palm oil mill effluent (pome), cocoa, peat and empty cocoa pods)	Restricted
By products of industries processing ingredients from organic agriculture	Restricted
MINERALS	
Basic slag	Restricted
Calcareous and magnesium rock	Restricted
Calcified seaweed	--
Calcium chloride	--
Calcium carbonate of network origin (chalk, limestone, gypsum and phosphate chalk)	--
Mineral potassium with low chlorine content (e.g. sulphate of potash, kainite, sylvinite, patenkali)	Restricted
Natural phosphates (e.g. Rock phosphates)	Restricted
Pulverised rock	Restricted
Sodium chloride	--
Trace elements (boron, Fe, Mn, molybdenum, Zn)	Restricted
Wood ash from untreated wood	Restricted
Potassium sulphate	Restricted
Magnesium sulphate (Epson salt)	--
Gypsum (calcium sulphate)	--
Stillage and stillage extract	Ammonium Stillage excluded
Aluminium calcium phosphate	Restricted

Sulphur	Restricted
Stone mill	Restricted
Clay (bentonite, perlite, zeolite)	--
Microbiological Prepararations	
Bacterial preparations (biofertilizers)	--
Biodynamic preparations	--
Plant preparations and botanical extracts	--
Vermiculate	--
Peat	--

" – " means decision yet to be taken. The conditions and the procedure for use shall be set by the certification programme.

"Factory" farming refers to industrial management systems that are heavily reliant on veterinary and feed inputs not permitted in organic agriculture.

APPENDIX 2

Products for Plant Pest and Disease Control

Certain products are allowed for use in organic agriculture for the control of pests and diseases in plant production. Such products should only be used when absolutely necessary and should be chosen taking the environmental impact into consideration.

Many of these products are restricted for use in organic production. In this appendix **"restricted"** means that the conditions and the procedure for use shall be set by the certification programme.

I. SUBSTANCES FROM PLANT AND ANIMAL ORIGIN	
Preparation of rotenone from Derris elliptica, Lonchocarpus, Thephrosia spp.	Restricted
Gelatine	--
Propolis	Restricted
Plant based extracts (e.g. neem, garlic, pongamia, etc.)	--
Preparation on basis of pyrethrins extracted from Chrysanthemum cinerariaefolium, containing possibly a synergist pyrethrum cinerafolium	Restricted
Preparation from Quassia amara	Restricted
Release of parasite predators of insect pests	Restricted
Preparation from Ryania species	Restricted
Lecithin	Restricted
Casein	--
Sea weeds, sea weed meal, sea weed extracts, Sea salt and salty water	Restricted
Extract from mushroom (Shitake fungus)	--
Extract from Chlorella	--
Fermented product from Aspergillus	Restricted
Natural acids (vinegar)	Restricted
II. MINERALS	
Chloride of lime/soda	Restricted
Clay (e.g. bentonite, perlite, vermiculite, zeolite)	

Copper salts/inorganic salts (Bordeaux mix, copper hydroxide, copper oxychloride) used as a fungicide maximum 8 kg per ha per year depending upon the crop and under the supervision of inspection and certification agency	Restricted
Mineral powders (stone meal, silicates)	Not allowed
Diatomaceous earth	--
Light mineral oils	Restricted
Permanganate of potash	Restricted
Lime sulphur (calcium polysulphide	Restricted
Silicates (sodium silicate, quartz)	Restricted
Sodium bicarbonate	--
Sulphur (as a fungicide, acaricide, repellent)	Restricted
III. MICROORGANISM/BIOCONTROL AGENTS	
Viral preparations (e.g., Granulosis viruses, Nuclear polyhydrosis, viruses etc.).	--
• Fungal preparations (e.g., Trichoderma spp. etc.)	
• Bacterial preparations (e.g., Bacillus spp etc.)	
• Parasites, predators and sterilized insects."	
IV. OTHERS	
Carbon dioxide and nitrogen gas	Restricted
Soft soap (potassium soap)	
Ethyl alcohol	Not allowed
Homeopathic and Ayurvedic preparations	--
Herbal and biodynamic preparations	--
V. TRAPS	
Physical methods (e.g., chromatic traps, mechanical traps, light traps, sticky traps)	--
Mulches, nets	--

"—" means decision yet to be taken. The conditions and the procedure for use shall be set by the certification programme.

APPENDIX 3

Criteria for the Evaluation of Additional Inputs to Organic Agriculture

Appendices 1 & 2 refer to products for fertilising of the soil and control of plant pest and diseases in organic agriculture. But there may well be other products which may be useful and appropriate for use in organic agriculture which may not fall under these headings. Appendix 3 outlines the procedure to evaluate other inputs into organic production.

The following checklist should be used for amending the permitted substance list for fertilising and soil conditioning purposes:

- The material is essential for achieving or maintaining soil fertility or to fulfil specific nutrient requirements, for specific soil-conditioning and rotation purposes which cannot be satisfied by the practices outlined in the standard or of other products included in Appendix 1. The ingredients are of plant, animal, microbial or mineral origin which may undergo the following processes:
- Physical (mechanical, thermal)
- Enzymatic
- Microbial (composting, digestion), and their use does not result in, or contribute to, unacceptable effects on, or contamination of, the environment, including soil organisms.

Their use has no unacceptable effect on the quality and safety of the final product.

The following checklist should be used for amending the permitted substance list for the purpose of plant disease or pest and weed control:

- The material is essential for the control of a harmful organism or a particular disease for which other biological, physical or plant breeding alternatives and/or effective management techniques are not available,

- The substances (active compound) should be plant, animal, microbial or mineral origin which may undergo the processes: physical, enzymatic and microbial and, their use does not result in, or contribute to, unacceptable effects on, or contamination of, the environment.
- Nature identical products such as pheromones, which are chemically synthesised, may be considered if the products are not available in sufficient quantities in their natural farm, provided that the conditions for their use do not directly or indirectly contribute to contamination of the environment or the product.

APPENDIX 4

USEFUL WEBSITES ON ORGANIC INPUTS

1. http://www.organic-research.com
2. http://www.cabi-bioscience.org
3. http://www.varanashi.com (Varanashi research foundation, India)
4. http://www.dpw.wau.nl/biob (biological farming systems research and teaching group, Wageningen)
5. http://www.hdra.org.uk/
6. http://www.fibl.ch (Research Institute on Organic Agriculture, Switzerland))
7. http://www.napcl.com (Nadukkara Agro Processing Co. Ltd)
8. http://www.apeda.com (Agricultural and Processed Food Products Export Development Authority, India)
9. http://www.andrewlorand.com (Consulting for Ecological & Biodynamic Agriculture)
10. http://www.phaladaagro.com (Phalada Agro Research Foundation Ltd, India)
11. http://members.shaw.ca/gardenab (Organic seeds and crops researcher)
12. http://www.howtocompost.org (Compost Resource)
13. http://www.erfindia.org (Eco-Science Research Foundation)
14. http://www.wormdigest.org
15. http://www.vasat.org
16. http://www.compostworld.com
17. http://www.pesticides.gov.uk (Pesticides Safety Directorate, U.K.)
18. http://www.vermico.com
19. http://cibrc.nic.in/ (Central Insecticide Board and Registration Committee)

20. http://members.shaw.ca/gardenab (Organic seeds and crops researcher)
21. http://www.aosa.org (Association of Seed Certifying Agencies)
22. http://www.seedanalysis.com/. (Commercial Seed Analysts of Canada Inc.)
23. http://www.dacnet.nic.in
24. http://www.agricoop.nic.in
25. http://www.agmarknet.nic.in
26. http://www.cibrc.nic.in

APPENDIX 5

RECORD KEEPING FORMS

Form-1

Compost/Pesticides Production Record A record of on-farm compost production practices.		
Farm Name or Unit: Production Year:		
Compost Pile, Windrow, or Unit I.D.: Date Started:		
Compost Production Method Used:		
Feedstocks Used (including inoculants):		Estimated C/N Ratio:
Dates	Temperature	Turned?

Form-2

Pest Control Activities and Inputs for Organic Crop Storage A record of the actions and materials you use to prevent/control pests in stored organic crops.		
Farm Name or Unit: Crop Year:		
Storage Unit I.D.:		Location (if off-farm):
Date	Pest Control Activity / Input	By Whom?

Index

A

Acetobacter 161
Agro-ecosystem 3
Antagonistic organisms 87
Application of mulch 14
Associative Symbiotic
Bacteria 161
Australian ladybird beetle 105
Azolla 73, 74, 162
Azospirillum 54, 64
Azotobacter 54, 63, 162

B

Banglore method 28
BGA 73, 75
Bio-Fungicides 88
Bio-Pesticides 88
Bioagents 97
Biocontrol Units 99
biofertilizer 159
Biofertilizer Production 51
Biofertilzer Production
Technology 52
Biomass production 11
Biopesticide Production 93
Blue Green Algae 75, 163
Bordeaux Mixture 125
Burning plant materials 21

C

Cattle-dung 127
Chilli-garlic extract 126
Chrysoperla 119
Chrysopid predetors 103
Compost 11
Compost heap 25
Compost Tea 35
Composting 23
Corcyra charging 101
Cover crops 5
Crop rotation 10
Cryptolaemus1 19
Culture environment 145
Curative Crop Protection
Methods 86

D

Diet preparation 108

E

Egg preparation 102

F

Fertility improvement 22
Furniture 40

G

Gram pod borer 110
Green manures 5

H

Human health 151

I

Indore method 26

M

Machinery 40
Microbial insecticides 87
Mineral insecticides 88
Mixed cropping 10

Mulching 13
Mycorrhiza 73

N

NADEP method 31
Nitrogen fixing plants 4
Non-Pesticidal Preparation 126
Non-symbiotic bacterium 162
NPV 106
Nursery Development 129
Nursery Technique 130
Nutrient recycling 19
Nutrient supply 19

O

On-farm resources 11
Organic Farm 49
Organic farming 155
Organic Input Evaluation 149
Organic manures 156
Organic matter 3
Organic standards 1

P

Pest Management 83
Pesticides 167
Pesticides Production Unit 119
Phosphate-Solubiliser 54
Phosphobacterin 66
Plant Disease 83
Plant Health 18
Plant nutrients 156
Plant Nutrition 18
Primary materials 23
Production of Vermicompost 41

R

Rhizobium 54, 60, 161

S

Seed stock 39
Sex pheromone traps 116
Soil compaction 17
Soil cultivation 15
Soil erosion 17
Soil Management 1
Soil-microcosm 3
Symbiotic association 161

T

Tissue Culture 132, 168
Tissue culture technology 141
Trichoderma 118
Trichoderma fungi 117
Trichogramma 119

V

VAM 165
VAT Method 29
Vermi-beds 39
Vermicast Collection 43
Vermicompost 12
Vermicomposting 37, 157
Vermiculture 44
Vermiwash 45
Vescular Arbuscular
Mycorrhizae 79

W

Water hyacinth compost 34
Water supply system 39
Worms 37

Z

zero-tillage 15